SYNC
MY
WORLD™

SYNC
MY WORLD™

A SEA & Nautical
Map to Relative Peace

Rodney St. Michael

SYNC MY WORLD™
A SEA & Nautical Map to Relative Peace

Cover Art, Inside Illustrations & Layout
by Rodney St.Michael

To order additional copies, contact:

Lulu Enterprises, Inc.
3131 RDU Center Drive, Suite 210
Morrisville, NC 27560
www.Lulu.com/rodney
sales@lulu.com
+1-919-459-5858

Printed in the United States of America

To the ASEAN Children

Your parents may be crazy,
but you don't have to be like them.

Chapter 1

Introduction

Introduction to General Convergence Theory

11-1-2003

When I was in Tokyo, Japan, several years ago with my colleague, I chanced upon an old officemate from Citibank, where I used to work. *It's a small world after all.* Even in other countries, it's easy to bump into an old acquaintance like Yashmine. As always, she was all-smile when she called my name and greeted me. She was with her mother, and we exchanged some stories about our recent lives. Everything seemed fine until her mom told me that she just had surgery in New York. She had a brain tumor removed, and her mother was glad that she remembered me since she was suffering from memory loss after the operation. I inquired about its cause, but they didn't know. There are suspicions, of course, on her heavy cell-phone use since she started using those electro-magnetic devices ever since they were still huge and heavy, unlike today's palm-sized low-radiation handsets.

It certainly gave me some insight on my own condition, schizophrenia. Although schizophrenics hear voices in their heads just like conversing with someone using a cellular phone or just like a 6-way conference call with your angry demanding clients from across the globe, most schizophrenics don't know how to control it or to turn it "off" during their first episode. Some sufferers also see "visions" just like the images that you see in your cell phone. Many of which are very disturbing, such as my first "vision," that I illustrated in my first book, and my second "vision," which I showed in the last chapter of my second book. Because of these "visions" and "voices," schizophrenics sometimes feel like they have "tumors" in their heads. And they go crazy. But in my case, over the years, I managed to harness the skill of turning it "off" at will through meditative techniques. It took a long time to develop it, but it was well worth the time.

Psychiatrists today are still continuously studying schizophrenia to determine its cause. They all have their own theories, and I have my own too. One of the more famous psychiatrists, Dr. E. Torrey, believes that it is caused by a virus. He cites other diseases such as Herpes Simplex, which has similar hallucinatory symptoms, to prove his point. He also mentions that many schizophrenics were fetuses during winter, and perhaps a flu-like virus in the mother's womb causes it. But today, most psychiatrists disagree with Torrey's theory since he seems to be describing diseases that mimic some schizophrenic symptoms but does not completely picture it.

Most of them believe that genes are responsible for it. Gottesman leads in this area. For example, in the case of the Nobel-prize winning scientist, Albert Einstein, his son was also schizophrenic, just like him. This is also true for another Nobel scientist, John Forbes Nash from the Oscar-winning movie, *A Beautiful Mind*, whose son is also schizophrenic. For a while, some psychiatrists thought that it is purely genetic, but relatively recent twin studies of parents who have schizophrenia did not indicate 100% occurrence of schizophrenia in identical twins, as it should have been if it were a purely genetic condition.

So now, doctors theorize that it is caused by a genetic *predisposition* toward it, combined with a significant element of a socially disruptive environment to trigger it. Last April, 2000 researchers in Toronto, Canada claimed that they have found the schizophrenia gene in the neighborhood of Chromosome 1. They were in hot pursuit. Two years later, in April, 2002, researchers in John Hopkins University in the United States announced that there were no schizophrenic genes in Chromosome 1. It was a false alarm, but they said they will try to look for it in Chromosomes 6 and 13, or even Chromosome 8. Admittedly, the researcher, Ann Pulver, says "It can't be explained by either a single altered gene or a single environmental cause." What is it with these genes, always playing peek-a-boo and hide-n-seek?

More recently, in July 2003, Nobel laureate Susumu Tonegawa of the Massachusetts Institute of Technology in the United States told the International Congress of Genetics in Melbourne, Australia that he *may* have discovered the schizophrenia gene. He claims to have created a genetically modified mouse *hypothetically* with schizophrenia. (Can a mouse tell you that it hears voices and sees visions?) The mouse was found to have "poor working memory—the kind of memory we need to find reading glasses we have put down, or our way back to a car in the carpark," according to the reporter. Tonegawa also said that his genetically modified mice don't like to sleep with each other, much like scientists exhibiting anti-social behavior. The "two mutants don't want to be together" when they sleep, Tonegawa says. The mice were also startled by loud noises, much like a cell phone ringing suddenly during a board meeting, which really upsets the board members. His research will eventually lead to new drugs, he says, to treat schizophrenia, so that he can patent the drug and make tons of money from people with mouse-brains.

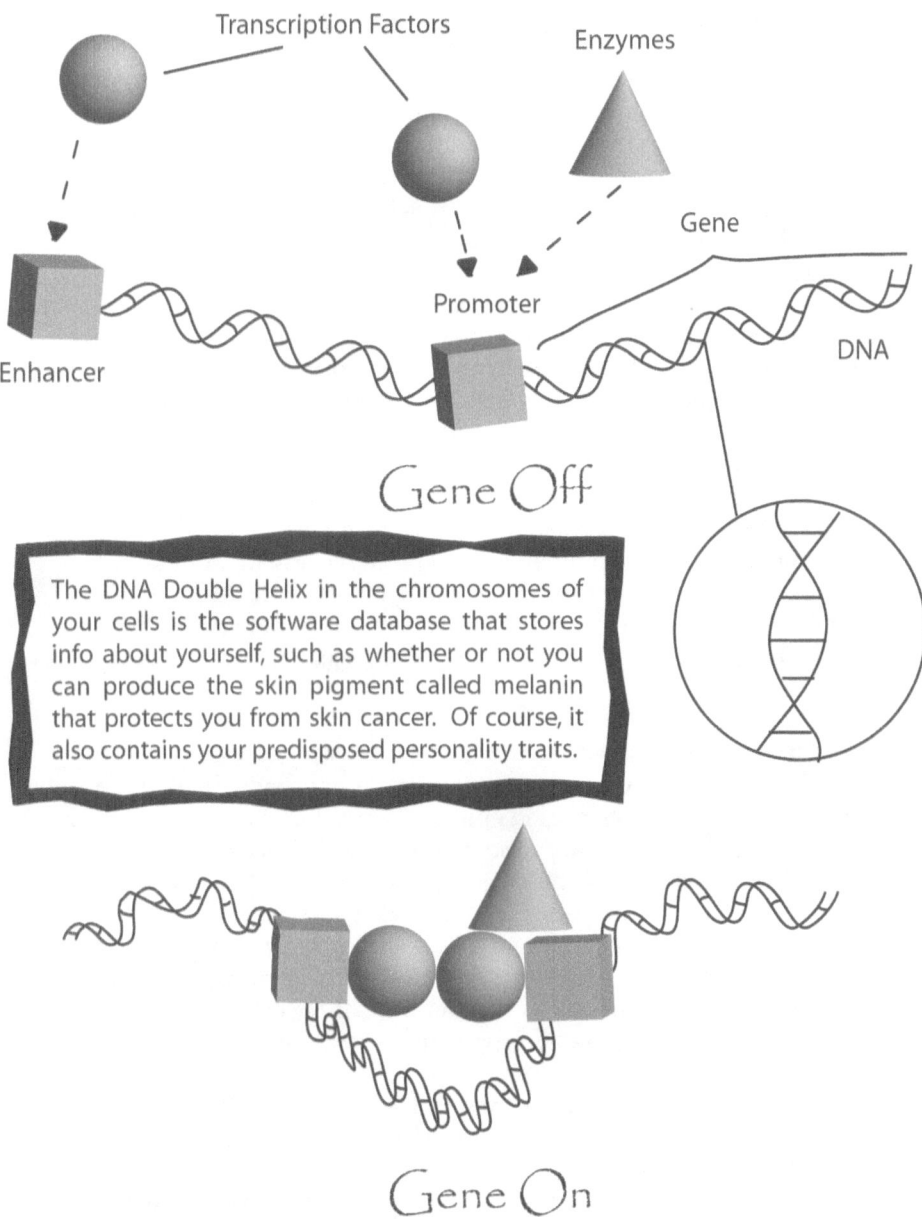

Transcription Factors

Enzymes

Gene

Enhancer

Promoter

DNA

Gene Off

The DNA Double Helix in the chromosomes of your cells is the software database that stores info about yourself, such as whether or not you can produce the skin pigment called melanin that protects you from skin cancer. Of course, it also contains your predisposed personality traits.

Gene On

Genes dynamically change all throughout your life by experience. But the most significant gene alterations occur from the time when you were a fetus until the age of six, when the brain fully develops. Certain genes become dormant or inactive if certain experiences do not occur.

To tell you frankly, Tonegawa's research seems to be the most absurd study that I have heard so far. Indeed, it will take some time before someone actually discovers valid gene data for schizophrenia. A schizophrenic can develop "poor working memory" if psychiatrists force them to take neuroleptics, as the case is in the United States. Neuroleptics of course belong to the same family of insecticides as Vx Nerve Gas, a Weapon of Mass Destruction (WMD) that George W. Bush is supposedly looking for in Iraq but is actually piled up in his own backyard in the United States. Of course, psychiatrists can also force patients to have "poor memory" through the hypnotic Power of Suggestion. A professional hypnotist can actually even give you Alien-hand syndrome, as proven by the Discovery Channel. But would I be able to write my first book if I had "poor memory?" No wonder the Discovery Channel also suspects fraud on the part of psychiatrists who claim of the existence of the rare condition which they call Multiple Personality Disorder (MPD) Syndrome, a disorder similar to amnesia, where one supposedly selectively "forgets" his other "selves" due to traumatic experiences, such as rape or priest molestation. This rare condition allegedly has only about 200 cases worldwide, unlike the common condition of schizophrenia which afflicts dozens of millions globally. And based on the psychiatric tapes of the famous MPD Sybil case, it seems to have been induced, as reported by the Discovery Channel.

Man, Oh man. All these "scientific," or more aptly called pseudo-scientific, attempts to discover something, clearly sheds the light on business rhetoric with the same spin as "scientific" statements that claim cigarettes and liquor are good for your health, just as insecticides are good for your brain. C'mon, guys, you can do better than that. Whatever happened to candor? Maybe all you scientists can stop playing with mice for a change and do something *normal*, like singing with a karaoke machine to the tune of Huey Lewis' paradoxical *Power of Love*. It all clearly points to the fact that all these "doctors" and "scientists" are crazier than their patients.

Perhaps the motivation for all these scientists should be revoked. Nobel-prize rhetoric paves the way to mad creations such as the nuclear bomb (thanks to Einstein) and game theory (thanks to Nash). Even supposed "treatments" for schizophrenia such as lobotomy (hammering the patient's frontal lobes with an ice pick) was awarded an enigmatic Nobel Prize. It makes you wonder what Alfred Nobel's true intent was in creating the Nobel Prize, since he is, after all, the inventor of dynamite. Is it to motivate scientists to create destructive tools as the Mensa Eugenicists strive for, or are they trying to produce more healing aids to cover up their own faults? Indeed, if they want to continue with this prize, why not focus it on peacemakers instead of scientists? Truly, that's why the *full* title of Charles Darwin's famous book, *The Origin of Species by Means of Natural Selection or the Preservation of Favoured Races in the Struggle for Life*, was "covered up" and called *The Origin of Species*. Isn't that also why Dr. David Kelly, the British WMD scientist ended his own life in 2003 after Prime Minister Tony Blair's government kept covering up their Iraqi war rhetoric? Maybe the Celts should start listening to Kelly's haunting *Voice* for a change.

It is interesting to see how the *Voices* of society's *five faces* produce madness. You only need to watch a few psychotic episodes of Warner Bros' animated *Batman* series, from Jack Warner's crew, to appreciate the haunting *Voices* of the *Scholar*, the *Shaman*, the *Business Class*, the *Militants* and the *Working Class*.

From the Scholars or *self-actualization class*, you have *Two-Face* (who holds the scale of justice and is a man in masquerade or a mask-wearer), *Riddler* (an enigmatic or paradoxical intellectual), *Ventriloquist* and his puppet *Scarface* (who has "Multiple Personality Disorder" with "Alien-hand syndrome") and *Mad Hatter* (a dark infatuated intellectual).

From the Shaman or *super-ego class*, you have *Harley Quinn* (a female psychiatrist and male-"ego" hater), *Dr. Emile Dorian* and *Man-Bat* (someone like Susumu Tonegawa), *Poison Ivy* (an extreme environmentalist and "green with envy"), *Clock King* (Six Sigma certified or an obsessive-compulsive perfectionist), *Mr. Freeze* (longs for "eternal life"), *Ra's Al Ghul* (a utopian society advocate), *Klarion* (develops "black magic"), *Cat Woman* (an extreme animal-rights activist), *Wormwood* (a trap plotter), *Clayface* (an obsessed actor who had plastic surgery), *The Joker* (a hebephrenic manic-depressive) and *Scarecrow* (a psychologist and fear monger).

From the Business Class or *social-needs class*, you have the *Penguin* (a high society business-class type but physically unattractive with low self-esteem), *Roland Dagget* (a businessman who creates diseases and antidotes for cash), *Grant Walker* (a businessman who prefers exclusivity and eugenics) and the *Terrible Trio* (a typical bored, rich group).

From the Militants or *security class*, you have *Red Claw* (a typical terrorist), *Gil Mason* (an archetypal crooked cop), *Maxie Zeus* (a classic corrupt customs agent or smuggler), *Lock Up* (a fanatical security expert) and *Bane* (an obsessed body builder using tons of "supplements").

And from the Working Class or *physiological-needs class*, you have *Farmer Brown* and *Amy Lou* (who produces genetically-modified agricultural products), *Killer Croc* (who sheds crocodile or fake tears like "wrestlers" who say "Pahingi ng pamasahe"), *Firefly* (who is unable to take rejection from the opposite sex and swears revenge), *Calendar Girl* (who vows vengeance on those who dropped her for younger girls), *Baby Doll* (who seeks retribution on her former employer) and *Sewer King* (a squatter enslaving his fellow poor).

Naturally, Batman is also insane, since he tries to study all of these "psychos," in his mountain cave, using his bat computer, continuously trying to help them recover, while he tries to live a schizophrenic life with a dual identity, under the mask of his bat-like radar ears, hoping to retire soon, with his teenage sidekick jumping around in his underwear, avoiding the mad house called Arkham Asylum, where his "enemies" live.

The Batman, of course, knows that all his "enemies" are actually part of himself, since everyone actually has a little bit of each "villain" within himself or herself. The "enemy" pops out only when you become imbalanced and extreme. Imbalanced people are actually quite common. For instance, many politicians and judges are like Two-Face. I also had an assistant once, who was like the *Joker,* and I encouraged him to learn about Eastern meditative techniques to help himself.

You will also easily spot preachers like *Killer Croc, Mr. Freeze,* and the *Mad Hatter.* In fact, after I published an essay last 11-1-2001 about clergymen, I was vindicated to hear soon after that about 542 cases of sexual abuse were filed against the Boston Catholic Church. Consequently, Cardinal Law resigned. And in other Catholic countries such as the Philippines, Bishop Yulung and Bacani also resigned after they were charged with sexual sins. (Manila's Cardinal Sin later left his office too due to health reasons, after they shed their crocodile tears.) Observe also how some female talk-show hosts are like Harley Quinn. And in the business world, you will see many *Penguins*, insecure and all made-up for the party.

Moreover, in the Catholic university where I graduated, I learned recently that an infatuated student, who faced rejection from a potential mate, stabbed his would-be lover like *Firefly*. (Perhaps the school administration should have listened to my advice.) The former *Clock King* Registrar of that school also falsified my thesis record after my graduation thesis there ranked No.1 for the trimester. It was only after I spread the word "lawsuit" that *Clock King* changed my grade back to its original score. The Green *Riddler* "Brothers," who run that split university (literally), also changed the name of their hotel, after identifying it with a "motel lord" and asking themselves, "What's in a name?"

Furthermore, many militants like communists and secret societies (Al Queda, Freemasons and the like) are like *Ra's Al Ghul*. The insane Western media also produced *Clayface* in Michael Jackson, just as they killed Princess Diana.

Of course, everyone knows that the American government is the *Ventriloquist*, throwing its voice on its *Scarface* puppet governments, like the Philippines. With its "multiple personalities," the fist minute it's a "Christian," and the next minute it's a high-society eugenics businessman, then as a mass-murderer, setting up "fake" bombings and using the Pearl-Harbor Gambit technique of fully knowing that an attack is coming, and yet allowing it to happen, to kill its own citizens, using it as a springboard and a moral cause for a eugenics attack on foreign soil, that earns them money at the same time, during the restoration period.

You can see the Ventriloquist-Scarface resemblance in George W. Bush and his puppet Gloria Macapagal-Arroyo, especially when Gloria visited Bush at the Whitehouse in 2003. The first "minute," George W leads an attack against Osama Bin Laden for the September 11 Tragedy. Then, when his government starts investigating the attacks, and learns that U.S. security agents actually knew of its imminence but ignored it, all of a sudden George W. censors several pages from the investigative report, which links the wealthy Bush family with the rich Bin Laden family, allegedly to protect Saudi Arabia, but is it perhaps, to protect the Bush family, who actually partnered with the Nazis during World War II, under George W's grandfather, Prescott, a director and co-owner of *Union Banking Corporation* (UBC), together with several Nazi executives?

UBC was seized by the U.S. government on October 20, 1942 under the *Trading with the Enemy Act* to halt Nazi banking operations in the U.S., under Prescott Bush. Then on October 28, the U.S. ordered the seizure of two companies under Bush and his partner Harriman— the *American Trading Corp.* and the *Seamless Steel Equipment Corp.* And on November 17, 1942, the *Silesian-American Corp.*, also under Prescott and his father-in-law, George H.W., was seized, but the U.S. government allowed the American partners to continue with their business. Even then, Hitler was already financed enough by the Bush family to continue with his genocide and eugenics propaganda. The Bush family would later, of course, move into the Central Intelligence Agency and later seat themselves as U.S. presidents.

Now, knowing all these, who is the "Real Bush?" Is it the one who's doing business with Bin Laden and the Nazis, financing or supplying them with arms and know-how against Bush's own country? Or is the "Real Bush" the macho cowboy who defeated his business partners by fooling the American people? Will the "Real Bush" stand up? *Scarface,* do you know the answer? Was it Hitler or was it his financiers who started World War II, in the interest of eugenics—the "science" of selective breeding, annihilation of races with "bad genes" and population control through *artificial* wars, famine, pestilence and disease? We do know that George W. Bush calls North Korea's Kim Jong Il as a "pygmy." Now that's something serious to think about.

Alas. In spite of the presence of all these "monstrous villains" in our daily lives, we do know that we can still live reasonably well with them. As I will explain later, this spectrum of "villainous" personalities has always been around since prehistoric times and will always be "around," since all people have two brains, a "male" left and a "female" right hemisphere, just like having two kidneys, two eyes, two ears, two breasts, two "balls," two hands and two feet. Unknown to many, everyone actually has a "split mind," but it can be managed through the technique called the Middle Way. You can actually easily remember this through the nursery rhyme:

> I have two brains, the left and the right.
> Hold them up high so clean and bright.
> Touch them softly, 1-2-3.
> Clean little brains are good for me.

Take note that you can actually "live" even with only "half" a brain, just as you can "live" with one kidney, in case you donate one to your relative. The same is true with "living" on one hand, one foot, or even one breast. Brain surgeons have actually removed "half" a brain from their patients, who managed to "live" with the other "half." But of course, life is not complete with only half of something, just as life would be "incomplete" without all these "villainous" personalities from Batman.

Most people often do not use their complete brains, but use only "half" of it. The left hemisphere of the brain controls the right side of the body, like your right hand. And the right hemisphere controls the left

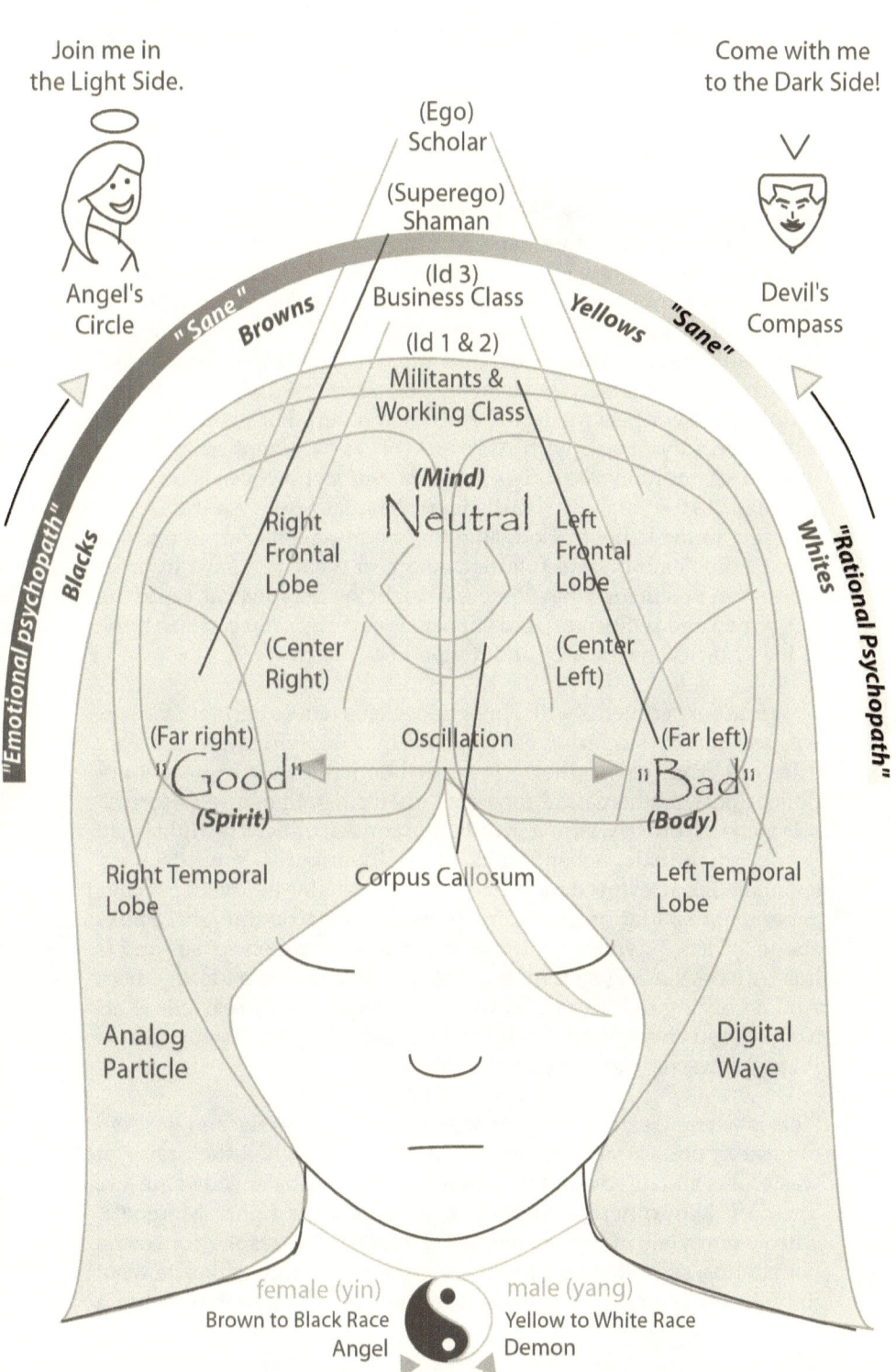

side. If all people were ambidextrous, they would use both brains, but most people are right handed. The division of the brains leads to "male" and "female" personalities, including the legion of persona that goes with it, depending on which part of the two brains become dominant. There is a "cable connection" between the two hemispheres. It consists of nerve fibers, the largest of which is the *corpus callosum*. And if the connection is damaged with lesions, you can get Alien-hand syndrome, where one hand has a life of its own and does something that your other hand wouldn't want to do. If the *corpus callosum* is cut entirely by a neurosurgeon, which is the procedure (a commissurotomy) done for patients with severe epilepsy, you will have a *split brain*. (Take note that the term *split brain* is different from *split mind*, which is associated with schizophrenics. It is also different from the term *split personality*, linked with the controversial MPD syndrome. People often confuse the terms with each other. For instance, some movies portray people with AHS or MPD as "schizophrenics." But those rare conditions are different. A *split mind*, however, is common and refers to the "inability" to distinguish fantasy from reality, usually referring to the belief in God, angels, demons, aliens, fairies and the like, or the "inability" to follow through in what you say. In other words, not practicing what you preach or being paradoxical, typifying clergymen and politicians, and varying in severity. You can also think of the *brain* as "hardware" and the *mind* as "software.")

Split brain experiments with epileptic patients whose *corpus callosum* was severed by surgery have revealed that the two brains are very different from each other. The left hemisphere is logical, two-dimensional, mathematical, language-oriented, grammatically-correct, self-inclined, orderly, strong in initiative and scientific. The right brain is creative, artistic, hebephrenic, musically-oriented, multi-tasking, spontaneous, dreamy, dance-inclined, emotionally-in-touch, psychic, superior in spatial or three-dimensional geometric thought, thinks about "others," gymnastically superior, has "common sense" and is holistic (sees the "big picture" and can assemble puzzle parts together). And according to Ehrenwald, left prototypes include Aristotle, Karl Marx, Freud and Apollo. On the other hand, right prototypes include Plato, Nietzsche, Carl Jung and Dionysus.

Here is where Western mad science verifies the accuracy and wisdom of Eastern philosophy. Neuroscientists have weighed the brains of westerners and discovered that their brains are imbalanced to the left. Their left hemisphere is literally heavier than the right. Moreover, when women talk more than men, their talkativeness results from using both brains, since the right brain alone *usually* cannot verbalize what it thinks of. But when a patient suffering from epilepsy, whose brain

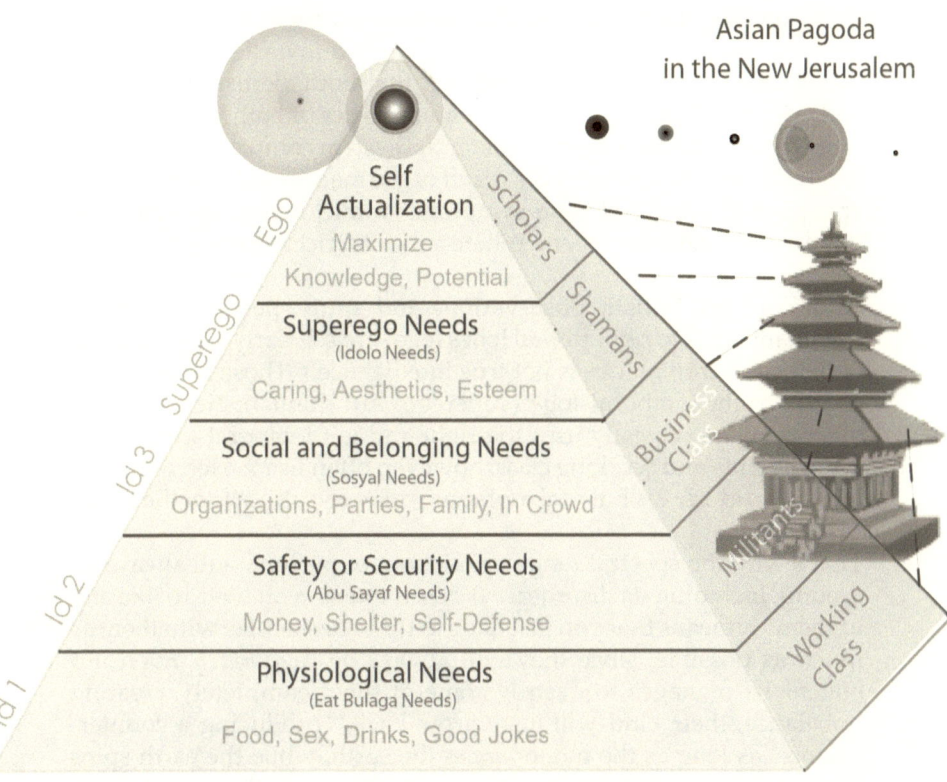

Asian Pagoda
in the New Jerusalem

Self
Actualization
Maximize
Knowledge, Potential

Superego Needs
(Idolo Needs)
Caring, Aesthetics, Esteem

Social and Belonging Needs
(Sosyal Needs)
Organizations, Parties, Family, In Crowd

Safety or Security Needs
(Abu Sayaf Needs)
Money, Shelter, Self-Defense

Physiological Needs
(Eat Bulaga Needs)
Food, Sex, Drinks, Good Jokes

Ego / Superego / Id 3 / Id 2 / Id 1

Scholars / Shamans / Business Class / Militants / Working Class

Asian Team Building
Retold by Abraham Maslow in 1954

$$\text{Satisfaction or Happiness Index} = \frac{PN + SSN + SBN + SN + SA}{5}$$

Where each need is ranked from 1 to 10, and a resulting index of 5 is normal

Autocracy

△

Ego on Top

Balance

✡

Harmonic Mix

ASEAN

Democracy

▽

Id on Top

Ben Franklin inserted these Asian masonic symbols on the back of the U.S. One-Dollar Bill when he was commissioned by the government to design it was split into two through neurosurgery, was interviewed, his *right* brain said that it wanted to be an automobile racer and that it *disliked* Richard Nixon, but his *left* said that it wanted to be a draftsman and that it *liked* Nixon; clearly indicating that both "leftist" Democrats and "rightist" Republicans have characteristics of both left and right, which makes the two-party presidential system confusing, deceptive, enigmatic and psychotic. (As a result of the madness of neurosurgery, epileptics develop AHS and symptoms mimicking MPD syndrome, thanks to the neurosurgeon. Where's the District Attorney?)

And from my holistic observations and study, people who have dominant left *and* right frontal lobes (ego) are scholarly types. (Using only the left frontal lobe is not true intelligence.) Those who usually use the right temporal lobe (superego) are shamanistic. And the business class typically uses the right and left temporal lobes (id3). The militants and working class, however, often use the left temporal lobe only (id 1 & 2). But schizophrenics may use all parts of the brain.

That's why the spectral range or rainbow of persona will always be around, including its disorders. It doesn't mean you have to like any of them. It means that you just have to try to harmonize with them as much as possible, since they will always be "a*round.*" Even if a eugenicist manages to destroy some of them completely, creating imbalance, their kind will just "grow back," producing a counter-balance, as long as the moon *circles* the earth, while the earth spins a*round* the sun, creating the four seasons, and as long as the sun *revolves* around the Milky Way, flushing itself toward the black hole at the *nirvana* center, like a giant *mixer*, evening out the cycle of personalities by modifying our liquid brain chemistries and genes, just as it changes the tides of the ocean.

This leads us to the definition of insanity. First, I will use the "wag the dog" definition from Lewis Caroll, the author of *Alice in Wonderland*, where you do the opposite of what is natural, such as smiling when you feel sad, or holding your lust when you feel sexual urges, or stopping yourself from bursting into anger when you feel mad, just like holding urination when you really feel like peeing. Of course you can do this temporarily, every now and then, without truly becoming nuts, just as you can hold your pee if you can't find a restroom. But if you do this regularly, just like stretching a rubber band to one extreme side, you'll eventually loose control and you'll become like one of Batman's "enemies." This is like a dog, which usually wags its tail when it's happy, that all of a sudden snaps and keeps barking, even when it's supposed to be happy and wagging its tail, because it has developed a general self-protection mechanism against the danger in its present environment. Usually, if the environment is safe already, it will regain control again, but the process can be expedited through the Middle Way.

Another way of explaining the difference between what is "sane" and "insane" is by observing a bull's-eye target and by using the archery term, *hamartia*. Here, "insane" or "abnormal" is anything extremely over and above, or lacking and below, or to the far left or far right of the relative center, or bull's eye. Being near the bull's eye then is "normal" or "sane." Subsequently, the terms "eccentric" and "centric" mean "abnormal" and "normal," respectively. But nowadays, these terms are often misused colloquially for spin or everyday rhetoric, especially when envy is involved, just as the word "ego" is often misused.

The mind is composed of three basic elements—"good," "bad" and "neutral" or the superego, id and ego, respectively. The id can be further subdivided into three sections—social belongingness, security and physiological. Or simply said, think of the mind as being composed of five people—a true scholar (ego), a true shaman (superego), a businessman (id 3), a militant (id 2) and a working-class man (id 1). When men talk about themselves and others think that it is the male "ego" or pride, for instance, they are really referring to the male "id 2" representing *security* needs, since it is the female that has the superego and the tendency to balance the left and right brains with the ego.

Likewise with the term "ego trip," which should be called "id trip" since it is due to insecurity. For example, some of the AXN action-series' contestants are there due to insecurity, but not all of them, since for some of them, it is a sport or hobby, which satisfies their true self.

Take note that most of the id is the left hemisphere or "bad" side of your brain. And the superego is frequently the right hemisphere or the "good" sacrificial-superhero Save-the-Whales side of yourself, while the "neutral" ego uses both sides of your brain.

Fists of Far Right

Fists of Far Left

The ego is actually silent most of the time, since if you are secure, you don't need to talk too much. But over the years, the id and the ego interchanged themselves, as a result of spin or rhetoric. Even in the East, in countries like India and China, the female "superego" is thought of as "bad"

Ashura from Nara Era, Japan

y = 3 Sin x

y = 2 Sin x

y = Sin x

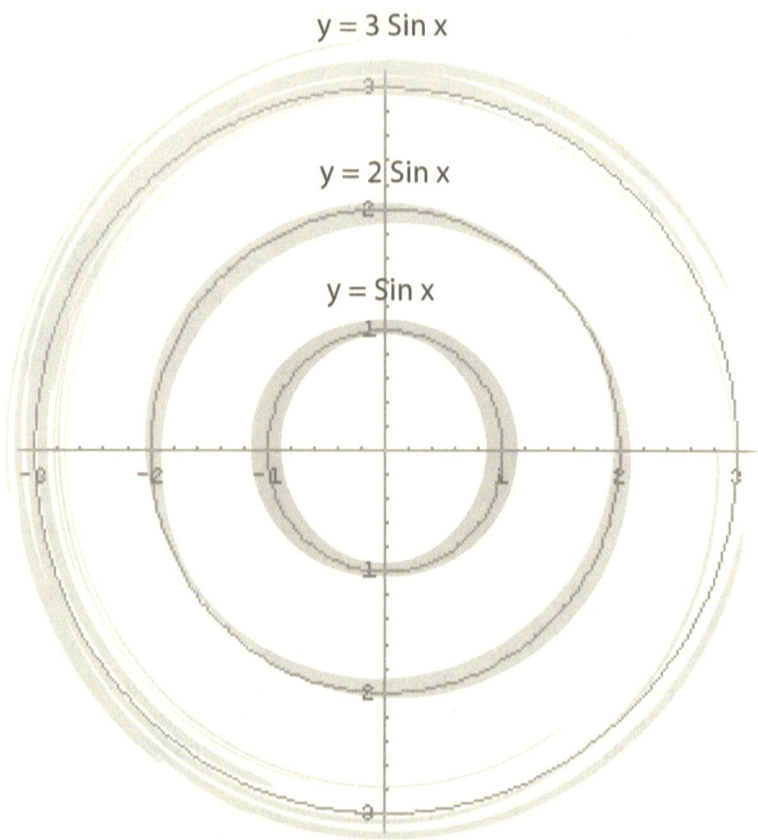

as a result of male-chauvinist rhetoric. But this is not the case in South East Asia, where the maternal society still exists. Take note also that legal expressions and documents are overly complicated nowadays because of too many balls of yarn spinning around to prevent rhetorical interpretation, which needs to be untangled.

Now that we have stopped the ball from spinning, we can now move back to the bull's eye and *hamartia* to define "insanity." (Actually, *both* sides of the ball are *true*. But I'll be explaining the light side of the ball first before I rotate it to the far side or dark side, to avoid confusion.) In Asian countries today, like Japan, a form of archery therapy called *kyudo* is still being practiced to this day. The archer just "let's go" and doesn't mind if the arrow doesn't hit its target. In a way it is like releasing yourself from the chains of Clock King. You can just keep on "aiming" and meditatively missing for thousands of times, but eventually, as time passes, you will hit the mark near the center, and be at one with yourself. Asian archers taught this system to their Greek friends, who used the word *hamartia*, which means "missing the mark at the center." The Greek word *hamartia* is actually found in the New Testament Bible, and English translators call it *sin*.

It is unfortunate today that most Christian clergymen mislead their flock, perhaps unintentionally, due to misunderstandings about the meaning of the word *sin*. *Sin*, as explained by the bull's eye analogy, simply means "off-center" or "eccentric." In other words, those who are far-left or far-right are "sinners" or psychopathic. This means that if a priest is celibate, he is actually being eccentric and therefore is committing *sin*. Of course, the same thing is true for the promiscuous Hindu guru in the Temple of *Kama Sutra*, who has "hundreds" of partners. They are both off-center or immoderate. In the same way, while the "rich" man is committing *sin*, the "poor" man is also a *sinner*. The case is also true if you are an "intellectual" or an "imbecile." And of course, a "war monger" and a "pacifist" do likewise. These two extremes tend to oscillate from one side to another, so that a criminal may become a priest later in life, and the preacher may become a murderer later, just like a rubber band stretched to one side and shoots toward the other side when the tension is released.

Take note that *sin* has nothing to do with being punished forever within a literal hell. Hell is actually the experience of suffering when we are always on the edge. And the cycle of suffering continues *perpetually* as long as we are immoderate. The experience of man in times of war, social conflict, pestilence, illness, famine and the like, is the idea of "hell." And this occurs when the Middle Way is ignored. Of course, as I explained in my first two books, there is also that possibility that others evolved before humans; that our minds are like software and our bodies are like computer hardware, and if your mind isn't "just right" when it is "uploaded" wirelessly to them, then it is simply sent to the Recycle Bin, where it may be "undeleted" and recycled, or where its contents may be "deleted" permanently. That's hell. But actually, for some, that's heaven. Indeed, to aim for either heaven or hell leads to failure. And to be truly "saved" is to be practical and "down to earth;" for if you try to "save" your life, it will be lost, just like trying to "delete" it. This is the essence of *salvation*.

Clearly, the one who is "just right," or generally moderate and centric, is the one who is "normal" and "sinless." In a way, it is like the children's tale of *Goldilocks and the Three Bears*, where she shuns the porridge that is too hot and too cold, eating the bear's meal that is "just right." *Repentance* then is simply turning around to prevent yourself from going over the edge, heading back toward the center.

Remember though that no one is like marks-man Robin Hood. That's why there is *grace* or *leeway*. For instance, this *breathing space* allows *latitude* for a young man to date several women first, before he selects a suitable mate. He'll never get it right the first time, and even if they settle with each other or if they choose to get married, their choice might not be correct, and they may decide to part their ways. But in the end, hopefully, they will find someone suitable for their brain chemistry and settle down completely, if they choose to do so.

Another feature of the arrow and the bull's eye is for navigation. Think of it like the "stick" of your manual-transmission vehicle and the driver's wheel. When you start your car, you shift from neutral to first gear. Then as you speed up, you shift to neutral and then to 2nd gear. You do the same thing for the other gears, shifting to neutral and on to the higher gears. But take note that you don't do this continuously. Sometimes, you slow down, and shift from a higher gear to a lower one. And although the driver's wheel is usually centered, since you are going straight most of the time, sometimes you will rotate the wheel to the left, sometimes to the right, and on rare occasions, you need to make a U-turn, or you need to reverse your direction, before you hit a dead end. Of course, the ancient Asians did not use a car for their navigation analogy. Instead, they used the mariner's or the sailor man's navigational tools, such as the *compass* and the *circle*.

Interestingly, the Chinese script for the word "China" is actually the script for the word "middle" or *tsong*. It looks like a rectangular ellipse or the sail of a Chinese *Junk* or ship, with a vertical line at the center,

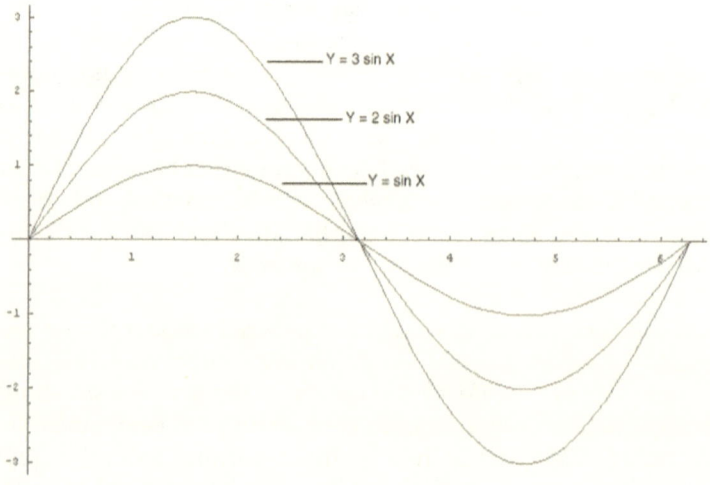

since China calls itself the "Middle Kingdom." (The term *Junk*, as it is used today to call trash, of course, comes from British rhetoric.) But ironically, China is not centered at the moment, since the working class and militant communists took over the country and murdered its scholars, shamans and business people, due to the insanity of foreign invasion. (The British navy destroyed the Chinese navy, during the Opium War, by pirating or learning from their "compass" and "circle.") Their security became too lax over the millennia as relative peace took over China after about 500 B.C. As a result, aided foreigners were able to use China's own technology against them, and consequently, the Middle Kingdom became crazy and shifted to the militant extreme. Although there seems to be a transition now from military rule to scholarly administration, the communist Chinese are still eccentric at the moment

This shows how the *compass*, which is used to draw the *circles* of the bull's eye paradigm, can also be used for dark or military purposes. If the Middle Way can make you sane, then pushing someone over the edge will drive him nuts. In a way, it is like pushing a ball, attached to a string (or a pendulum), to one extreme side. Releasing it will then make the pendulum oscillate to the other side, and swing to and fro, with a high amplitude (or height) at first, but it will eventually "calm down" gradually, as the amplitude decreases, toward the middle.

In trigonometry, this pendulum cycle can be translated to sine waves. A *simple* sine wave has the equation $Y = Sin\ X$. The term "*sin*" for this curve is not a coincidence since many ancient mathematicians are eccentric. A sine wave can be translated into a circle, where higher amplitudes mean larger circles and shorter amplitudes translate to smaller circles, homing in on the bull's eye. $Y = 3\ SIN\ X$, for instance is larger or taller than $Y = 2\ SIN\ X$ and $Y = SIN\ X$. So now, you can answer the question, "whY do we experience suffering and "hell"?" Because of great SIN. And "whY do we experience happiness and "heaven?" Because of moderated SIN. Naturally, if you are totally sinless, your heart-rate monitor is probably moving in a straight line already, and you have passed away from this earthly existence, reaching the peace of *nirvana*.

These cycles, waves or circles can illustrate the cycles of war and peace, life and death, business booms and busts, night and day, and the four seasons—winter, spring, summer and fall. Of course, it can also be used for other types of cycles such as the growth cycle, which is used by bankers, businessmen and even artists or entertainers for their products, services and organizations. A product, service or an actor, for instance has a life cycle that grows, peaks and then declines.

Another example is the cycle of learning. If you study through traditional schooling, graduating from college might already be the optimal way that you can benefit from this learning method, since if you continue with Masters or Doctoral degrees, you may end up becoming dumber than undergraduates, or even elementary school children. This is referred to as the *Law of Diminishing Returns*.

To prevent this from happening, re-invention, or creating a new growth cycle is needed, when you reach the peak of your current cycle. This is where the dark side or "insanity" comes in, since this method is often used in business and military strategy. "Normally," countries or regions have cycles with small amplitudes. It is not really "normal" to have true peace, for instance, since the more peaceful a nation is, the more it will get bored, and it will tend to move toward war. The same is true for a nation at war. It will tend to move toward peace. In other words, pushing a nation to extreme utopian peace will actually create war, and vice-versa. Another example is the economic cycles. Pushing a nation's economy to extreme booms will produce depressions. In a way, it is like creating manic-depression and schizophrenia within a nation.

Let us use a "hypothetical" example to see how it is applied by the "great" nations. Suppose that I buddy-up with other people and create an "underground" society called the *Jolly Roger Society*. These techniques will be taught to the members, depending on how much "evil" capacities they have. The more "evil" the initiate, the more knowledge he will be given. But the "kinder" he is, the lesser the know-how revealed to him. The members will be chosen from the "rich" and "powerful" crowd. One member, for instance will be from X Family, owning X bank and other businesses. If I knew that the cycles of boom and bust were heading toward bust because I pushed the economy previously to a roaring boom, I would have to produce an anti-cycle policy to prevent this from happening.

In fact, I could create a new growth cycle, through a few dirty tricks. I would look for countries with the ability to pay huge loans, and I would search for an extremist in that country—someone perhaps who got a bad Charlie-Chaplin haircut from a "Jew" and later maintained a *square* mustache to symbolize this. This *square-cut* extremist would have written a book sharing his utopian dream for the nation. I would then fund the development of this man, hopefully lifting him up to the office of Chancellor, or even as a Great Dictator.

As soon as he swings to this extreme, he will develop mad delusions of supremacy, and he will try to round up all those people who gave him a *square cut*, throwing them to the neuroleptic nerve-gas chambers. He will even declare war against the neighboring countries that pissed him off during what could have been his "normal" years. Then his entire region will be at war, perhaps even a Great World War. Of course, by the time the war ends, their region will need *loans* and *business supplies* from me and my businesses, creating a *new growth cycle*.

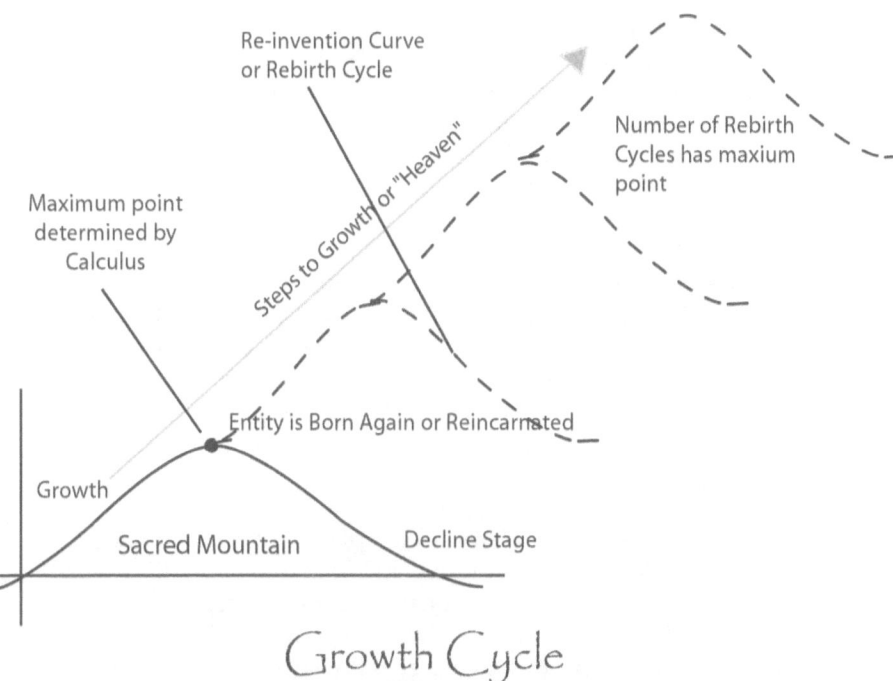

Growth Cycle

Everything has a beginning and an end, but sometimes people try to push growth to new heights through re-invention. Sometimes this is possible, but in other cases, it becomes unethical and breeds detrimental consequences. A car, for example, has a useful life of about ten years, just like marriage, but on the fifth year, it starts to experience problems and then its usefulness declines. If you regularly refurbish it though, like an old Volkswagen Beetle, its useful life can be extended. This is true for companies, products, artists, countries, relationships, careers and so forth, which is why ancient Asians have always been studying the management of cycles. And since nothing lasts forever, you never really own anything. Hence, managers try to weigh the pros and cons of renting and purchasing, whether it be in relationships, equipment or land.

But even before that happens, my own country can declare war against the Great Dictator's country, thereby filling up orders for my steel mills and other businesses, which supply my country's military. This would be the *Jolly Roger's* anti-cycle policy.

Hypothetically, I could re-use this policy, or rather, recycle this *modus operandi*, over and over again, since most people are currently unaware of it anyway. I could "do business," for instance, with Y and Z Family in the Middle East, push them over the edge, make them extremists, supply them with arms, and make tons of money in the process. Then, if they strike back at me, I will declare them as "crazies." And when they start plotting attacks against my country, I will allow it to happen, and use it as a springboard or moral cause to attack Y and Z's nations, or where their extremists are hiding, thereby testing the new military equipment of my own country. I will then create a puppet government after I destroy Y and Z. Subsequently, I will earn money again when I reconstruct it and provide it with loans.

I could even perhaps incite world anger against a "pygmy" in the Far East, causing a chain reaction that will wipe out the Clock Kings and gain more money in reconstructing their Scarred Faces, as part of my anti-cycle policy for the "greater good" of mankind. I would be terribly mad, but no one would notice it, except perhaps, the *Union Jack Society*. So, I will also try to obliterate them before they reveal the fact that X = Y + Z.

With all these *lefties* running around the planet, plotting to kill each other, who would be *left* in the world? There is actually a very interesting version of *Camelot*, starring Richard Gere, called *First Knight*. It is, of course, about the classic love triangle between King Arthur, Lancelot and Lady Guinevere. It is also about the Knights of the Round Table and the rebel knight who used to be a part of it, but who eventually turned to the *dark side*. In the movie, King Arthur invited Lancelot to be one of his knights, and Arthur asked Lance which door would he choose, left or right, in case the castle was in danger, and if they needed a hasty way to escape and live. Lancelot said that he didn't know. In this case, he actually chose *right*. But Arthur told him that he should choose *left*, since it was the way to survive. Ironically, Arthur died in the end. Although he was *center-left* (the *left* ego), his *far-left* (id) rebel knight killed him, leaving Lancelot (the *right* ego) and Guinevere (the superego) surviving together in the *right* center. (This is usually also true for right-minded women, who outlive left-minded men, since men often kill each other and leave their widows behind. Women have 10% more neurons in their brains than men, even though men are usually bigger and have larger brains. The

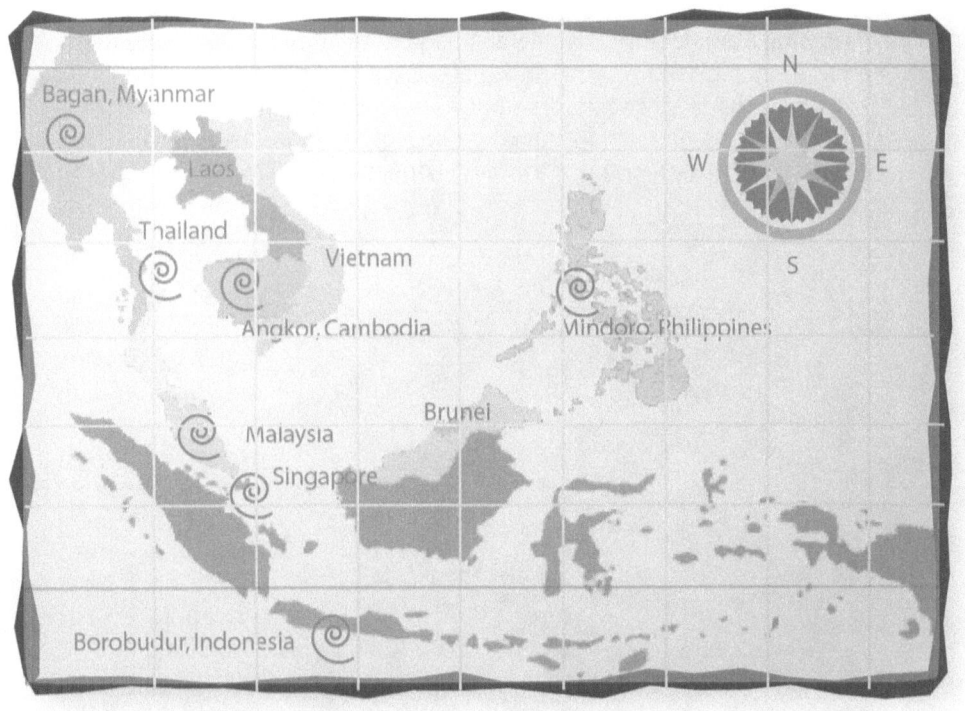

The Association of South East Asian Nations (ASEAN)

corpus callosum—the "cable connection" between the two brains—is also larger in females. Hence, women tend to use both brains, while men usually use only "half" a brain—the left hemisphere.)

This has led me to conclude that the 10-nation region called the *Association of South East Asian Nations* (ASEAN) is the most "normal" of all regions in the world, since the U.S., the U.K., Australia, Spain, and Italy are approximately *mid-to-far left* (id 1-2-3); Canada, Russia, the European Union (EU) in *general*, Scandinavia, Japan, South Korea, and the soon-to-be China are roughly *center-left* (left ego); the ASEAN is around *center-right* (right ego); and Tibet, South-Central Asia, the Middle East and Africa are more or less *mid-to-far right* (superego). (Take note again that because of political rhetoric and security-reflexes, what is called "right" and "left" is often confusingly interchanged, since "left" or "wrong" is also "right," and vice-versa.) You can also think of the U.S. and its "gang" as the rebel knights; the EU and its comrades as the *Knights of the Round Table* with the dead King Arthur, rising up from the grave, as China; and the Lone-Ranger ASEAN as Lancelot, with Tibet and her Little Lulus as Lady Guinevere.

The central Stupa doesn't contain an image of the Buddha indicating that he has freed himself from the prison of suffering because he is no longer oscillating. His heart monitor is now a straight line and he has attained *nirvana*, perhaps graduating into the ranks of the gods, freed or maybe issued parole by the warden of the earth correctional facility.

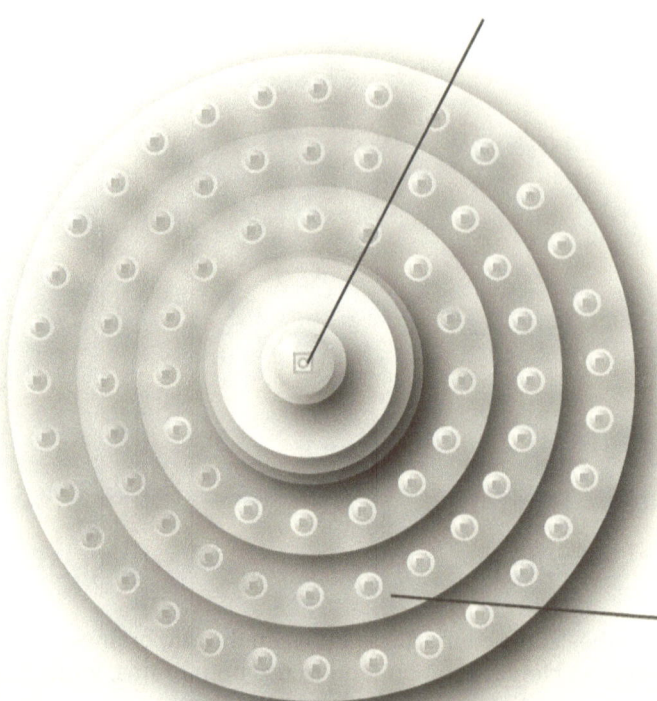

Each of the 72 stupas cage an image of the Buddha indicating greater suffering at the extreme cycles or outer rings. Moderate "sin" symbolizing the inner rings indicates less suffering or "imprisonment." It shows that the Middle Way leads to a happier life, so shift your gears properly.

Stick-Shift Apex of Borobudur Temple Complex

This temple is the best Buddhist complex in the world. The reliefs in the temple complex show the life of the Buddha for pilgrims to study. It also shows the journey of Sudhana, the hero of the *Gandavyuha,* in his search for truth. His story is about the impermanence of a bubble, reminding you that nothing lasts forever and that the cycles must be managed to produce happiness. The secret basement under the temple was originally hidden from public view until some Japanese soldiers during World War II uncovered it. It reveals carvings of Indonesian everyday life and also some gruesome images of suffering. It symbolizes the hellish existence of dwelling too much on the id. The temple is therefore divided into three: (1) Id at the basement (2) superego hero carvings as you climb toward the top (3) and an ego apex.

Unlike the unsightly giant golden buddhas of Thailand's Wat Po, which are iditistic, and unlike the fat Chinese laughing buddhas, which are also idististic representations of Chinese monks, the life-sized Javanese buddhas of Borobudur display the Middle Way most appropriately. Indeed they are the best buddhas in the world and are most fitting to the Javanese.

The 72 bell-shaped stupas is another representation of yin-yang. It is the male and female genitalia joined together in union. If you peep inside the cage bells, you will see a life-sized Javanese buddha, with varying hand gestures depending on the stupa; some are teaching, others are meditating or doing another task to manage or control the imprisonment of yin-yang upon himself. You might not be able to control your race, gender, genetic predispositions and the like, but you can manage it to a certain extent through the Middle Way.

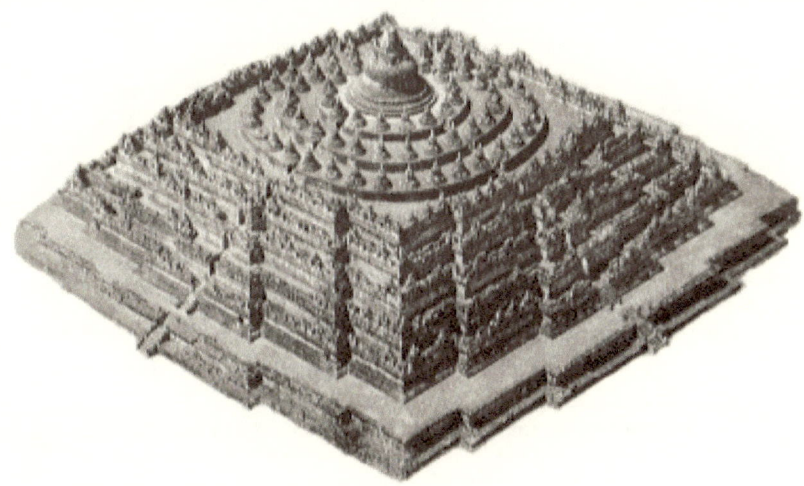

This pyramid buddhist temple was built more than 1,200 years ago by the people of Java, Indonesia. A trip to Bali and Borobudur works well.

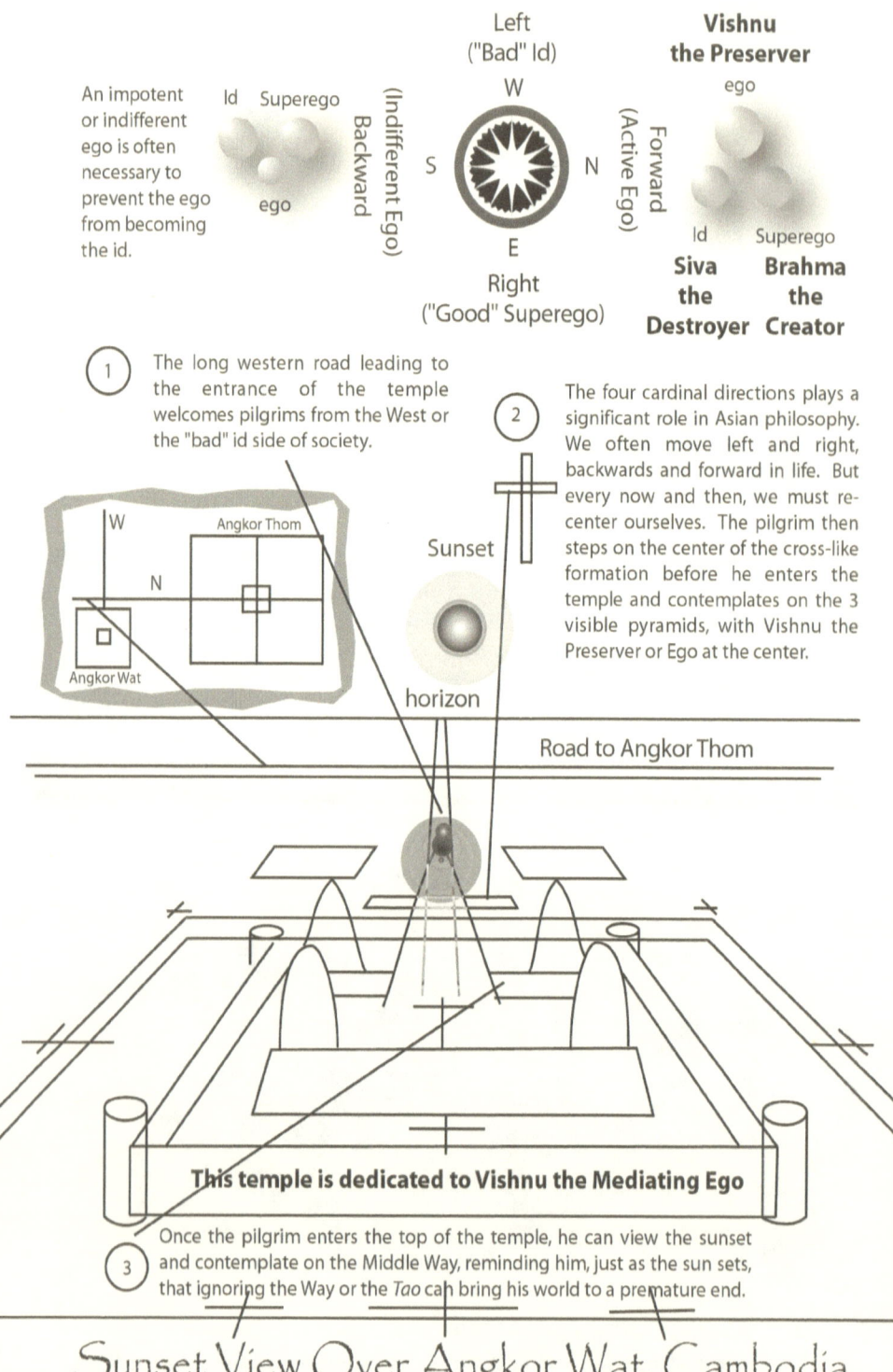

Left
("Bad" Id)
W

**Vishnu
the Preserver**

ego

An impotent
or indifferent
ego is often
necessary to
prevent the ego
from becoming
the id.

Id   Superego

(Indifferent Ego)
Backward

ego

S

N

Forward
(Active Ego)

Id   Superego

**Siva
the
Destroyer**

**Brahma
the
Creator**

E

Right
("Good" Superego)

① The long western road leading to the entrance of the temple welcomes pilgrims from the West or the "bad" id side of society.

② The four cardinal directions plays a significant role in Asian philosophy. We often move left and right, backwards and forward in life. But every now and then, we must re-center ourselves. The pilgrim then steps on the center of the cross-like formation before he enters the temple and contemplates on the 3 visible pyramids, with Vishnu the Preserver or Ego at the center.

W
Angkor Thom
N
Angkor Wat

Sunset

horizon

Road to Angkor Thom

This temple is dedicated to Vishnu the Mediating Ego

③ Once the pilgrim enters the top of the temple, he can view the sunset and contemplate on the Middle Way, reminding him, just as the sun sets, that ignoring the Way or the *Tao* can bring his world to a premature end.

Sunset View Over Angkor Wat, Cambodia

By taking note also of tangible data and intangible experience, I noticed that the West is currently too "rich," and Africa is too "poor." Scandinavia and Canada seems to be too "peaceful," and the Middle East and Africa seems to be too "war torn." Japan is like Clock King and China is still recovering. Korea, of course, is still like Two-Face, thanks to the Ventriloquist, staging another anti-cycle strategy to prevent King Arthur from rising like a Phoenix. But in the end, it will be very sad indeed to see any of Batman's "enemies" go, and leave the rest of the world as weeping widows. Truly, all and sundry have to play their role for everyone to learn, since as Shakespeare said, "the whole world is a stage," and "fair is foul, and foul is fair," but since we already know the ending, can't we all change it, and say that we care?

So, what is the "secret" of the ASEAN? Observe its regional map, and you will notice that at the far-left of the region are militant countries such as Myanmar. At center-left is Buddhist *Thailand*. At the right and far right are the Roman Catholic Philippines and Muslim Indonesia, respectively. But at the *center* of the ASEAN is *Singapore* and *Malaysia*. Malaysia is a peculiar nation, since it is racially and spiritually heterogeneous, yet is still homogeneous when their diverse citizens relate to each other. Half of the Malaysians are Muslim Malay, a third is Buddhist-Confucianist-Taoist Chinese, and the minorities are Hindu Indians and Christians. It is very interesting to see how tame Malaysian Muslims are, relative to other heterogeneous countries such

as the ones in Eastern Europe, the Middle East, Africa, and even the far-right of the ASEAN, in the southern Philippines, where Christians clash with Muslims regularly. This led me to conclude that the Buddhist-Confucianist-Taoist philosophy of the Malay-Chinese is really what can tame the "beast" within us all. The combination of the three "ways" neutralizes its acidic and basic qualities. And to use it is good medicine for the world, not just Asia.

Interestingly, it is in the /\SE/\N region where Hindu-Buddhist Bull's-Eye Temples reside, as described by Plato in his description of /\tlantis. And this is also why this region has historically been raped and beaten by Westerners for hundreds of years. But as resilient as it is, it has managed to survive, although still very sick at the moment, especially the countries at the far left and far right. The European influence of Marxism has maimed it. The forced acceptance of Christianity in the Philippines has made it psychotic, and has dubbed this country as the "Sick Man of Asia." Lastly, the Islamic influence of the Arabs on the southern Malays has made it mentally unstable. Remarkably though, their cheerful happy-go-lucky nature is still intact, even if they are dressed up in foreign "clothes." To dress it up again in its true "clothing," or rather, to undress it back to its original self, will take a long, long time, but through Taoism's *do-nothing* philosophy, I'm sure that it will recover and heal itself eventually within this *millennium*, if the earth is still around.

This led me to the realization that the best theory for explaining and healing schizophrenia, bipolar disorder and other similar conditions is in Asian 5-element Divinology, led by the Taoist arts, "expedited" through the Buddhist Middle Way's *Noble Eightfold Path*, and managed by Master Kung Fu's (Confucian) delivery system—the *Eight Steps of Learning*, or what Westerners now refer to as *The Fifth (V) Discipline*. And in the future, I believe that more people from around the world, not just in Asia, will learn and use these theories to make themselves happy and fulfilled. Collectively, I call all of these, including the South East Asian *Nautical Map to Relative Peace* that I will lay down later, as the *General Convergence Theory*.

Let us now remember the various theories for the cause of schizophrenia, as laid down by psychiatrists and geneticists. One theory says that it is caused by a virus. But since schizophrenia occurs around the world almost *evenly* (about 1% of the population), it makes you wonder how a virus could spread so evenly worldwide. Is there a giant mixer that distributes viruses evenly worldwide? If so, then why are viruses such as HIV so unequally scattered globally? HIV-positive individuals are mostly in Africa and India, and some are in other parts of Asia. This makes the schizophrenia virus theory weak, and probably gives more strength to the HIV euro-eugenics theory.

The other theory laid down by geneticists is the genetic-predisposition-plus-social-environment theory. However, even genetic disorders that share the same theory do not have even distributions around the world. For example, Europeans have a high frequency of *cystic fibrosis*. In fact, most Britons carry the gene, but *do not necessarily develop* the disease. Africans, on the other hand, have a high occurrence of *sickle-cell disease*. And Asians have a high incidence of *thalassemia*. People can actually carry a copy of the gene but they don't necessarily have to have the full-blown condition. Even multi-gene disorders such as *cancer* and *hypertension* do not have an even distribution around the world because of varying genetic make-ups and environment among different races and countries. But if schizophrenia is evenly distributed around the world, regardless of race or nationality, it would cast doubt on the second theory. Another strange thing about this theory is the fact that *everyone* has some type of genetic disorder. That means that everyone would need gene therapy, including the eugenicists, since all of Batman's "enemies" come from all walks of life—the Five Faces of Society. But what would mankind be if that were done? Robots?

There are also some European psychiatric theorists who think that a mother's age during pregnancy may have something to do with the development of schizophrenia. Doctors know that the risk for fetal abnormalities increases as the mother ages. For example, a 20-year-old mother has a 1 in 1,500 chance of having a baby with Down's syndrome. But at 45, her chances are 1 in 35. (Indeed, some men consider women to be "over the hill" when they reach 27—the age when the menopause cycle begins and having a baby becomes riskier and less likely.) But in the case of schizophrenia, the connection between the mother's age and the child's mental condition is weak; although it could play a minor factor. In my case, my mother was 25 at the time of my birth.

Moreover, some parents report that vaccines contribute in some way to mental illness. For instance, they claim that the vaccine Measles-Mumps-Rubella (MMR) can cause autism. And other types of vaccines have been linked to other types of conditions. Dr Mark Geier, from the Genetic Centers of America, linked 133 *reports* of neurological problems and brain damage with MMR. Of course, these were only *reports* from parents posted on the American Vaccine Adverse Events Reporting System (VAERS) from 1994 to 2000, and the ones who are allergic to these vaccines are the most susceptible to it. Interestingly, when I was a foreign student in California, U.S. law required me to

Electromagnetic Spectrum

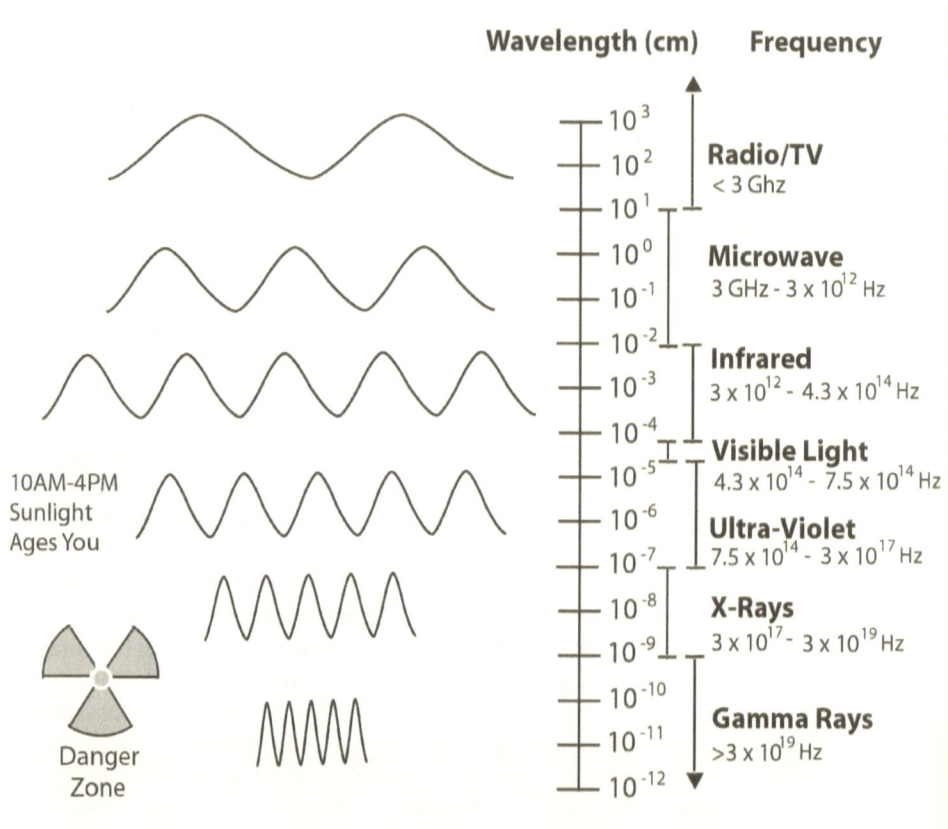

Wavelength (cm) **Frequency**

10^3	
10^2	**Radio/TV** < 3 Ghz
10^1	
10^0	**Microwave**
10^{-1}	3 GHz - 3 x 10^{12} Hz
10^{-2}	**Infrared**
10^{-3}	3 x 10^{12} - 4.3 x 10^{14} Hz
10^{-4}	**Visible Light**
10^{-5}	4.3 x 10^{14} - 7.5 x 10^{14} Hz
10^{-6}	**Ultra-Violet**
10^{-7}	7.5 x 10^{14} - 3 x 10^{17} Hz
10^{-8}	**X-Rays**
10^{-9}	3 x 10^{17} - 3 x 10^{19} Hz
10^{-10}	
10^{-11}	**Gamma Rays** >3 x 10^{19} Hz
10^{-12}	

10AM-4PM
Sunlight
Ages You

Danger
Zone

Radio Wave Spectrum

100KHz	1MHz	10 MHz	100MHz	1GHz	10GHz	100GHz
	AM Radio	Shortwave Radio	TV	FM Radio	Cell Phone	Microwave
VLF	LF	MF	HF	VHF	UHF	SHF	EHF

Visible Light Spectrum

Many animals see in black and white. but apes and fishes see in color.

Red	Orange	Yellow	Green	Blue	Violet
drive, direction, energy, courage	vitality, fun	creativity, clarity	renewal	relaxation	harmony

take three consecutive shots for MMR, tetanus and polio in 1993. And several weeks after I took the shots, I developed paranoid schizophrenia. Dr. Torrey may be right to a certain degree. It is indeed possible that it plays a contributory factor in some way, although definitely not entirely since most people who take these shots do not acquire autism or schizophrenia.

Take note also that if you have schizophrenic *symptoms* and you use illegal drugs such as LSD, you will develop hallucinations, but your condition is *not* schizophrenia. It's simply called *drug addiction*, although many people confuse it often with the former. That's why illegal drug abuse is *not* a cause of schizophrenia.

Hence, there must be another theory that could more accurately explain schizophrenia completely, or at the very least, supplement the other theories. It has something to do with a giant mixer that circles the globe evenly. Ironically, this theory—which I used to think of as pseudo-science, since scientists often overlook it, because of impressions of magical crystal balls and the chain of control and bondage—is actually a more reasonable theory after all. Ironically, it is the theory that actually allows you to manage your own destiny. And after several years of studying schizophrenia, I have concluded that the best theory for explaining it and the range of people's personalities, comes from the cyclical influence of gravitational and electromagnetic waves from the universe, which molds fetal formation and genes during the nine-month pregnancy period. It produces a brain-chemical imbalance that predisposes the child to various types of conditions and personalities, including schizophrenia and bipolar disorder. Then upon exposure to non-harmonious environments, which is also influenced by the universal cycles, schizophrenia may develop. The imbalance, of course, occurs in all people, and it varies from person to person, depending on the season when the child was conceived. In fact, Torrey believed that schizophrenia is virally caused because he noticed the numerous cases of people who had it and who were born during winter. Thus, he thought that it might have been a virus such as the Flu virus that caused it. But even in equatorial regions, where there is no winter, and where I was born, it still occurs.

Hence, the astrologers have something going for them after all. And the gravitational wave influence of the moon, sun, planets, plus the electromagnetic influence coming from various areas of outer space, does have a *partial* influence in shaping who we are now. And in all likelihood, it also shaped the evolution of various human races—including all the genetic disorders that we carry to this day—and the spectrum of flora and fauna in planet earth.

Albert Einstein's *Gravitational Theory* or the *General Theory of Relativity* explains the varying influences of gravitational waves as bodies of mass spin around each other. The moon, for instance, rotates around the earth, producing the moon cycles, from new moon to full moon. Then the earth revolves around the sun, creating the four seasons, as the sun circles the black hole in the center of our Milky Way. These cycles cause the tides to change from hi to low. And since your brain is composed of two-thirds water, and perhaps even more while you swam in your mother's womb when you were a fetus, the gravitational wave cycles that vary from heavenly bodies, as you developed in the womb, produced your predisposed personality, along with your genes, which was also shaped by the cosmic cycles thousands of years ago. And just as everyone's finger prints or palm lines are unique, nobody has exactly the same brain as anyone else.

The ancients linked this phenomenon to the zodiacal constellations, thinking that it had something to do with a mystical force that produced 12 basic types of personalities. The sun's relative position among the constellations in the zodiac created popular names for these personalities, but the stars itself in those constellations have no real influence, contrary to the opinion of popular fortunetellers. It has more to do with the earth's relative position vis-à-vis the sun, which varies gravitational wave influence. This means that the ancients were able to plan the predisposed personality of their child by timing their copulation to produce a scholar, shaman, business-class type, militant or working-class type. And although the West used a solar calendar, the East used a more accurate *lunar* calendar to build their civilization, using the heavens as a guide. Hence, insanity is also called *lunacy*.

Moreover, electromagnetic wave influence on fetal formation is known by obstetricians and radiologists today as a reality. For instance, radiologists forbid pregnant women to be exposed to *X-rays* since it might lead to fetal abnormalities. An obstetrician might also warn a mother-to-be against exposing herself to *microwaves* from her oven. Some scientists also think that living near powerful *radio-wave* transmitters causes cancer or tumors. And prolonged exposure to a television screen or computer may lead to myopia (nearsightedness), cataracts, or at the very least, dizziness and headaches. Even popular science fiction has exaggerated this in the form of comic strip heroes such as Stan Lee's *Incredible Hulk*, wherein Dr. David Banner is exposed to *gamma rays*. And whenever he becomes angry, he becomes a green monster.

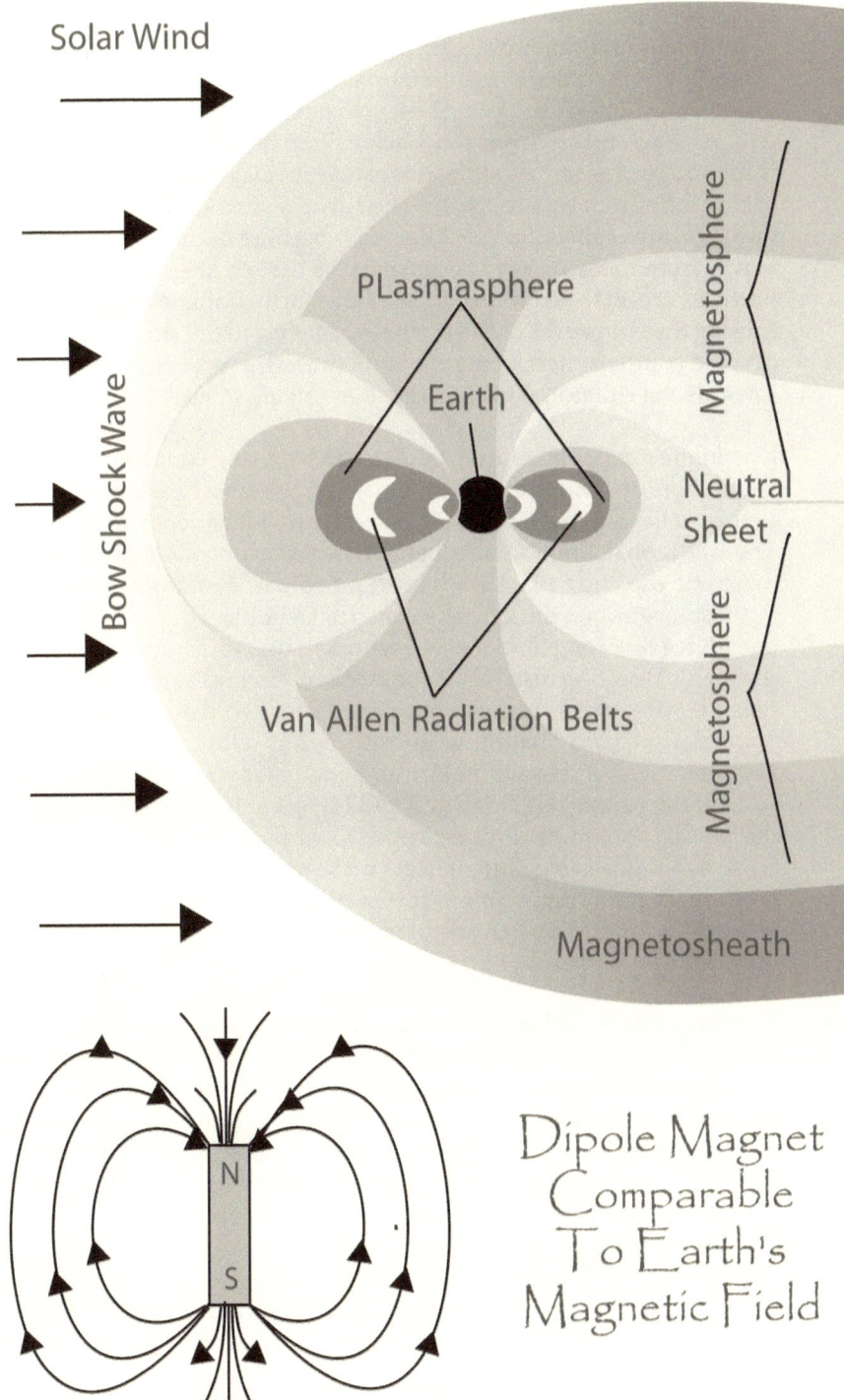

Dipole Magnet
Comparable
To Earth's
Magnetic Field

Remember, though, that all these types of waves are bombarded to us continuously from outer space, in varying degrees, depending on the season or the location of the earth, relative to the sun. For instance, *cosmic rays* (although not electromagnetic) come from the sun and exploding stars called supernovae. The sun also emits *gamma rays*, *x-rays* and *ultra-violet rays*. These waves travel to the earth, together with protons and electrons in what is called the solar wind. Much of it is filtered by the earth's atmosphere and its magnetic field, but some of it still finds its way to earth. The earth is surrounded by a deadly radiation area called the Van Allen belts because of these particles and waves from the sun. And because of the earth's magnetosphere, the particles are attracted to the poles of the earth, both north and south, causing the *Aurora Borealis* (Northern Lights) and the *Aurora Australis* (for the southern hemisphere). Sometimes the particles are thrown towards the equator. This is called a *magnetic storm*.

It is interesting to note that traditional Japanese couples time their copulation or sexual intercourse during the onset of the Northern Lights, believing that their children will turn out better as a result of it. The traditional Chinese also believe in the art of *Feng Shui*, where magnetic direction plays an important role in avoiding misfortune. For instance, when the Chinese construct a house, they will first seek the advice of a Feng Shui master, who will use his compass to survey the area. Direction also has its equivalent yin and yang. One of the reasons, of course, for this practice is to avoid locations where "poltergeists" and "haunting ghosts" occur. Obviously, the earth's magnetic field, along with electromagnetic waves coming from the sun, rotating along its 27-day axis and 11-year solar-activity cycle, plus the predisposition of certain people towards it causes "hauntings." So, to avoid this, Feng Shui masters make sure that your apartment or house is in harmony with the forces of nature. (Be wary though of obsessive-compulsive Feng Shui masters. Too much of them is not a good thing.)

Scientists also postulate that these waves can change the weather on earth, which may indirectly affect the fetus through the mother's mood changes. In fact, there is a psychiatric disorder called Seasonal Affective Disorder (SAD). People are generally gloomy during the winter and livelier during spring or summer, which is caused by the tilt of the earth's axis and its relative location along its elliptical orbit around the sun. Combining this phenomenon with solar activity will definitely affect the fetus.

Radio waves, on the other hand, come from various heavenly bodies, such as stars, comets, planets, galaxies and clouds of gas and dust.

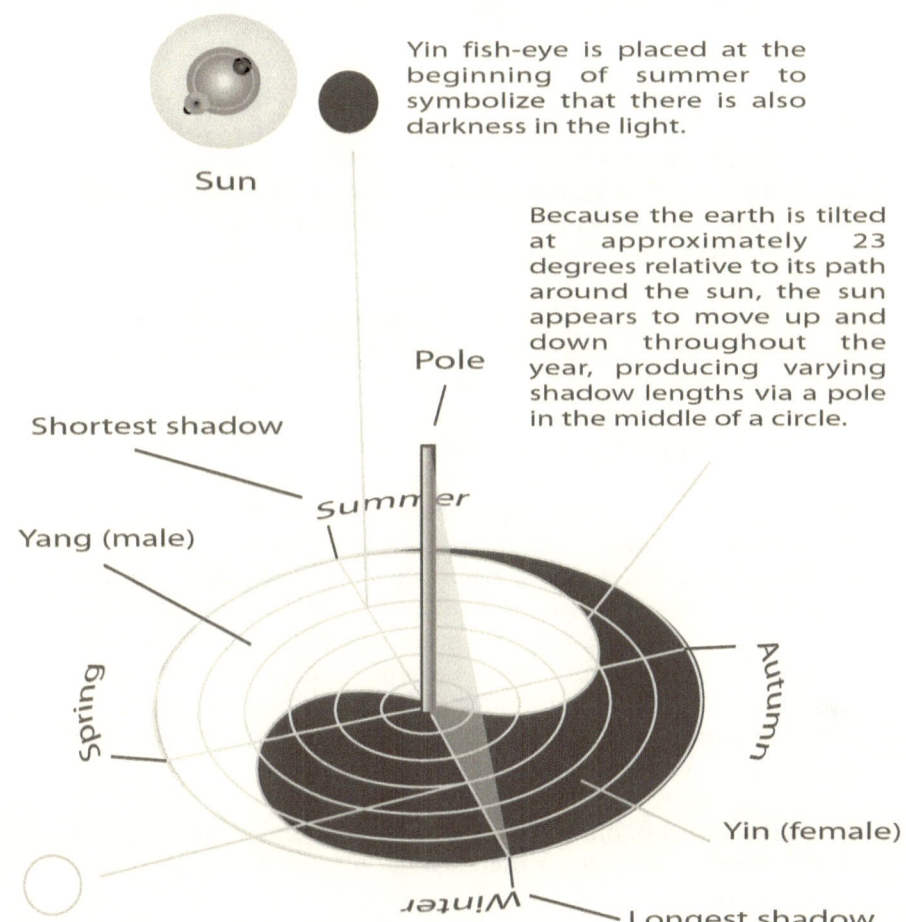

Yin fish-eye is placed at the beginning of summer to symbolize that there is also darkness in the light.

Sun

Because the earth is tilted at approximately 23 degrees relative to its path around the sun, the sun appears to move up and down throughout the year, producing varying shadow lengths via a pole in the middle of a circle.

Pole

Shortest shadow

Summer

Yang (male)

Spring

Autumn

Yin (female)

Winter

Longest shadow

Yang fish-eye is placed at the beginning of winter to symbolize that there is also light in the darkness

Origin of Yin-Yang Symbol

Ancient Chinese astronomers developed the Yin-Yang symbol, as a representation of Change Management, by observing that everything in the Universe is part of a spectrum, with its most basic form in either two states (digital) or three states (analog) depending on whether you look at it from the male (two) or female (three) perspective.

The intensity of these waves varies, depending on the object, and its wavelength can measure as long as a football field or as short as a basketball.

This has occurred, naturally, since prehistoric times, shaping the evolution of man. For instance, if certain types of electromagnetic waves hit an animal's DNA, its offspring may become an albino or white mutant. This is seen in animals such as white elephants, which are revered in Asia, since south-east Asians consider abnormalities as forms of "good luck." But many types of animals can become albinos too. It also occurs in humans and this has evolved to races such as the white man or *Caucasian* with blue, green or grey eyes and blonde or strawberry hair. This type of mutation produced different abnormalities such as a pointed nose, which can't smell as well as a round nose; and white skin, which is prone to skin cancer since it is deficient in *melanin*, normally found in normal "colored" people. Albino eyes are also sensitive to light waves and their nervous system and genetic brain chemistry is predisposed to a militancy which some might believe is near to that of the *Incredible Hulk*. These people would later be called "pirates." Due to their nature, the "evil" Africans, who were "normal," kicked them out from where the original humans lived. The "pirates" or Caucasians eventually had to migrate to cooler, icy areas in Europe, where they were able to adapt better because of their abnormal condition. But the "pirates" vowed vengeance, just like *Mr. Freeze*, and said, "I'll be back!" Indeed in the future, he would say, "Haste la vista, baby!"

On the other hand, other people were bombarded by cosmological waves, together with aging parents and racial mixing, which produced offspring similar to that of Down's syndrome, but *not quite*. These "Mongoloids" moved to the northern Far East. Other races also developed through sexual relations between the races. Caucasians, Mongoloids and Africans produced various types of brown people, tall and short, such as the Indians, Malays, Australian aboriginals and Native Americans, since some people find certain abnormalities, such as dimpled faces, white skin, blue eyes and blonde hair to be attractive. Thus, natural selection included abnormalities.

Take note that when nature produces a spectrum, it tends to converge toward the center. In the case of the human-race spectrum, black was produced first, followed by white, followed by their mixture, the Asians or the Third Race, which can relate to all races since everyone is a part of them. In other words, the spectrum or rainbow of humanity, including the races and their personalities, evolved through "Astrology."

Ancient Asian scholars noticed that the convergence toward the center is what produces eventual peace and stability after a chaotic evolutionary beginning. Indeed, it was easy for them to develop the Buddhist Middle-Way philosophy that everyone—"good" and "evil"— are all parts of the same whole.

Asian scholars studied this evolutionary phenomenon, thousands of years ago, and documented the effects of the "mystery forces" in books such as the *I-Ching* (*Book of Cycles or Changes*) and in *Tzu Ping* forecasting (Four Pillars of Destiny). They used it like a meteorological or weather guide to prepare themselves for the "weather" ahead, as they navigated through the ocean of life. Unlike the rain or storm, which is clearly visible in the sky, gravitational waves and electromagnetic waves are invisible (except for light), and would need indirect charts to recognize their presence.

Through astronomy, ancient Asian "astrologers" can produce a *5-element chart* that shows your brain-chemical predisposition. (Until recently, this knowledge was kept secret from the Asian commoner. Only the Asian elite were able to use it.) These elements are linked to your level of predisposed needs: [1] *physiological needs* (working class), [2] *security needs* (militants), [3] *social or belongingness needs* (business class), [4] *super ego needs* (shaman class), and [5] *self-actualization needs* (scholar class), as retold by Abraham Maslow to the West in the Hierarchy-of-Needs pyramid.

These five elements will determine whether you are predisposed to becoming one of the five faces of society—shamanistic or artistic class, scholarly or intellectual class, business or sociable class, militant or security class, or working/unfulfilled-potential class. Some people will be a combination of different class types, but many people will not reach their full potential. They usually drop to the working class due to natural tendencies among humans, such as crab mentality and the Law of Supply and Demand. Thus, the hierarchy becomes triangular instead of being equal.

Your 5-element chart tells you what your need levels are. If your scholar-class and working-class needs, for instance, are high, and all the rest is low, then you must work out a plan to balance yourself by concentrating on the three other low elements that needs to move up, in order for you to "normalize." Your brain usually automatically does this, without your knowing, by trying to pursue the needed activities to balance yourself, such as the pursuit of a career in the area or areas where you are lacking to harmonize your brain-chemical imbalance. Hence, the low elements are called by Chinese medicine men as "lucky" since it will make you happy and complete.

European doctors have also experimented with the fetal-formation theory using Western Astrology. In the 1950s, Stanislav Grof, a Czechoslovakian Psychiatrist, experimented with the Astrological birth charts of schizophrenic patients, and he garnered positive results. As a prescription then, for this condition, he advised a form of therapy using super-oxygenated yoga techniques, similar to the practice of ancient Taoists, Buddhists and Hindu gurus for healing themselves. Grof's research led to the development of a new branch of psychology called *Transpersonal Psychology*.

The ancient Asians, of course, used *breathing, meditation* and *exercises* such as the *Tibetan Five Rites*, as the keys to balancing out brain chemistry. They use it for imbalanced *Qi* or *Chakras*. And a very simply way to do it is to go out in a lush garden at night, when the plants start to release oxygen. Then sit in a cross-legged position, close your eyes, take a deep breath through your nose, hold it for a few seconds, then exale. Do this several times while focusing on your breath.

This breathing exercise is also a form of *Qi Gong* and can be practiced also along the beach, since the saline (salty) water molecules in the air, will contribute to better breathing. You can also inhale the steam of boiling salted water in your kitchen if you don't live near the sea or ocean. People who have mental or emotional disorders usually have allergies that clog up their respiration. The exercise is a good way to heal both your allergies and emotions, just as swimming can also be used to strengthen your breathing.

And even though your brain is only about 2% of your body weight, it uses up 20% of your oxygen supply. It only takes 3-5 minutes of oxygen deprivation for your brain cells to die. Disease also proliferates if your cells do not have sufficient oxygen. That's why after a period of studying or reading, your head starts to feel tired, indicating the need for rest to allow enough oxygen to refuel your brain. Blocked nasal passages due to allergies will also weaken your brain because of insufficient oxygen. But through breathing exercises such as yoga, Qi Gong and Buddhist meditation, you will be able to oxygenate your brain and balance out its chemicals.

You can also supplement this with hydrotherapy to remove the toxins in your bloodstream coming from imbalanced brain chemicals and histamine. If you dip yourself in a pool, for instance, you will notice your hands pruning like a raisin after a few minutes. This is caused by osmosis, where water and toxins in your body seep in or out of your skin's pores, depending on the pool's and your body's pH level. Take a dip in a hot spring or in beach water; never chlorinated swimming-pool or tap water. (Traveling to South East Asia or the Pacific Islands is ideal, but it can also be simulated in a good spa or wellness center in your local area.) You may also take a diuretic-type of tea, such as *Banaba*, to flush out toxins through your urine. Then drink at least five to eight glasses of filtered water daily to re-hydrate your body. Never drink tap water since it may contain heavy metals, nitrates, radioactive compounds, chlorine, fluoride, bacteria, parasites or petrochemicals. (People who drink and bathe in chlorinated water also have a 50% greater chance of acquiring cancer in their lifetime. You also absorb six times more contaminants through a ten-minute shower than through all the water that you drink in a day. But a shower filter will solve this problem. The filter quality or brand will determine the amount of contaminants removed. At the moment, a Japanese microwater [antioxidant water ionizer] machine is recommended, but this may be unaffordable to many people, so try to find the best type that fits your budget.) Take note that dehydration can cause mental problems and allergies.

This is probably why Pisceans, like me, are attracted to water, the color aqua-marine and seafood. There seems to be an automatic mechanism in the brain that tries to balance itself out intuitively, just like autonomic responses such as sneezing or coughing, as an immune reaction to respiratory infection. Fishes in sea water are also our ancient evolutionary ancestors, and your body needs to take a dip in the sea to be at one with its roots.

Moreover, healthy cells are neutral, fed by a well-balanced, least-processed diet. It should neither be too basic nor too acidic. For example, you will feel sicker the more acidic your cells are. If the cell pH is below 7.0, your cells are diseased. Consuming a diet laden with red meat will increase the acidity of your body. On the other hand, fish or seafood, together with rice, fruits and vegetables is good for you. In fact, "primitive" neutral diets rich in minerals and activators of primate Asia-Pacific peoples have been linked to their immunity towards tooth decay and their overall healthy systems that greatly defend against degenerative diseases *prior* to the introduction of modern processed foods to their diet. Basic elements include calcium, sodium, magnesium and potassium. On the other hand, acidic elements include chlorine, sulfur and phosphorus. Stress, improper rest, excessive exercise and pollution will also increase your body's acidity—sometimes to the point of borrowing calcium, a neutralizer or alkaline-forming element from your bones, to maintain normal pH levels. (Calcium deprivation then leads to osteoporosis or dwarfing of the skeletal system as you age. It makes seniors appear shorter.) This is why it is important to balance out or neutralize your system.

By balancing out brain chemicals or hormones, you can also heal a variety of other diseases and conditions, including aging. Eating Disorders, for instance, is an emotional disorder resulting from imbalance. And if your metabolism is fast, like Clock King, you will probably develop a form of Anxiety Disorder, making you appear scrawny or thin. On the other hand, slow metabolism will make you fat. Balancing your five elements may even help in slowing down the onset of AIDS in an HIV-infected person, or it may even help cancer patients, since meditative yoga is proven to increase your anti-bodies and to strengthen the immune system.

Your sleeping postures and vision can also tell you something about your imbalanced behavior or brain chemistry. When I was in California ten years ago, I started to wear eyeglasses. The Californian doctor said that my myopia and astigmatism was due to genetics. But I found out later after several years that just like psychiatry, ophthalmology and optometry was also laden with medical fraud. It turns out that myopia or nearsightedness is caused by near-point stress—the practice of habitually focusing on near objects such as in reading or computer use instead of focusing regularly on distance objects, as in the case of our ancient ancestors who used their eyes for hunting. And it turns out that astigmatism is caused by bad posture. For instance, if you are always seated at the extreme left or right side of a classroom, you have to tilt your head and focus your eyes awkwardly toward the teacher at the center. If you do this constantly, you'll develop astigmatism. The same is true for reading while in bed, or reading side-wise while laying down your body, or any other imbalanced posture. Now, of course, after doing some eye exercises and correcting my bad habits, my vision has significantly improved. In a few more months, it might become 20-20 again. I visited an optometrist lately and she was surprised that my vision improved so significantly. Apparently, many eye doctors are still in the dark or are intentionally keeping people in the dark to maintain their livelihoods in destroying people's eyes since "corrective" lenses will blind you over the years. (Synonymously, the Far Left and Far Right are astigmatic, whereas the Moderates have clearer vision.)

Try to observe also how you sleep. Some people sleep in a left-sided fetal position. Some sleep on their right side. Others sleep straight on their backs, and still others sleep on their tummies. If you sleep on you left side, brain fluid and blood will flow more to your left brain, under-nourishing your right brain. So try to balance it out by sleeping more often on your right. Then center yourself. Do the same thing for other positions. Try to counterbalance it. Also avoid beds that are too soft, such as water beds. Ancient Asians slept on hard surfaces, and that is usually best for your back.

Consult a licensed physician first, knowledgeable in *both* Western and Eastern medicine, before you do any of the suggested exercises to balance your system. As a rule of thumb, Western medicine usually gives quick relief against symptoms, since Westerners are usually impatient, but the long-term results of the treatment are destructive. On the other hand, Eastern medicine is usually slow, and it requires discipline to follow the prescription, such as the exercises, but the long-term results are better since it may give you an actual cure instead of simply relief. (Yoga actually means "yoke" or a "discipline" that you carry upon yourself for self-improvement.) Thus, use Western approaches for convenience or emergencies, but use Eastern techniques in the long-run. For instance, if you have *allergic rhinitis*, and you are taking anti-histamines for temporary relief, it will be good to follow the eastern prescriptions too for a long-term permanent cure. Eventually, you won't need Western medicine anymore to balance your system.

However, many times you are unable to balance yourself due to social road blocks and hindrances. Most people, for instance, are not aware of this system, and the various class types of society will produce conflict against each other, which will often contaminate you in the process. Thus, there is a need to teach this system to others too, so that they can harmonize with others, like you.

The Christian Europeans during the Dark Ages were hard to teach and convince though. Many of the loony eccentrics of Europe back then actually "discovered" many "new" things as a result of trying to prove to the authorities that "Astrology" was actually reasonable. For instance, Johannes Kepler (1571-1630), the German astronomer who produced the *laws of planetary motion* that describe *elliptical* planetary orbits, in reality produced horoscopes for a living. The Polish Astronomer, Nicolaus Copernicus, who discovered that the planets revolved around the sun, instead of the earth, did not publish his work until a few days before his death, for fear of the Christian authorities. But the Italian astronomer Galileo Galilei (1564-1642) proved the work of Copernicus and was later locked up, under house arrest, for his discoveries that disproved the authority of the Church. While imprisoned, he continued his work, trying to figure out the nature of the *Force* behind Astrology, which led him to study *gravity* by using weighted balls.

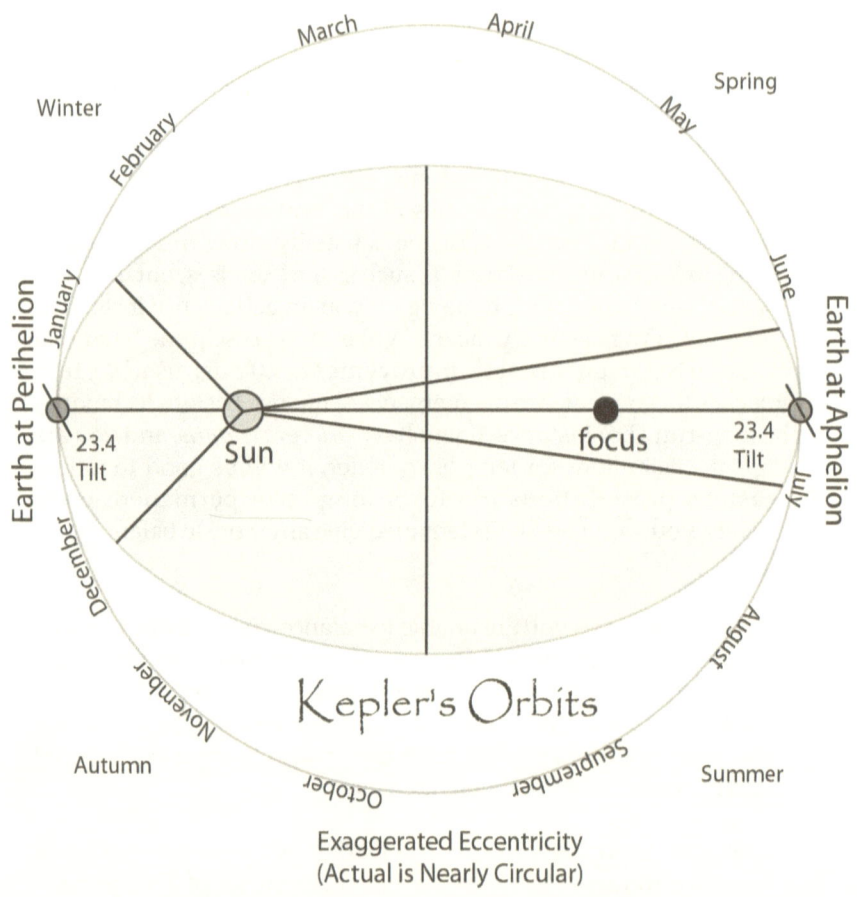

Kepler's Orbits

Exaggerated Eccentricity
(Actual is Nearly Circular)

Date of Perihelion moves 1 day every 58 years.
About 10,500 years ago, perihelion was during the summer

The British physicist and astronomer, Isaac Newton, continued Galileo's work and formulated the *law of gravity*: $F = G M_1 M_2 / d^2$ (where F is the Force between two masses M_1 and M_2, such as the mass of the earth and the moon, d is the distance between the masses, and G is the universal gravitational constant). Take note that Kepler's laws of motion say that planetary orbits are *elliptical* or *eccentric*, not circular. The planets are also tilted on its axis. Newton's Law then says that the greater the distance between the masses, the smaller the *Force*. Therefore, the *Force* will vary between, let's say, the earth and the sun, as the earth revolves around the sun. In my case, as a Piscean, when I was conceived in late Spring, the *Force* was weak with me, but it became stronger as my birth neared, finally becoming a little weaker when I was born during late winter. This weak-strong-semi-strong combination created me. (This of course is an over-simplified explanation, since Newton's Law is imperfect, and there are many other masses, such as the moon and other planets, involved.)

It was in 1665, during the bubonic plague in Europe, that Newton formulated his greatest discoveries. Because of the plague, the school where he was studying, Trinity College, shut down. And since he became free from the fallacious, packaged-Aristotlean knowledge from

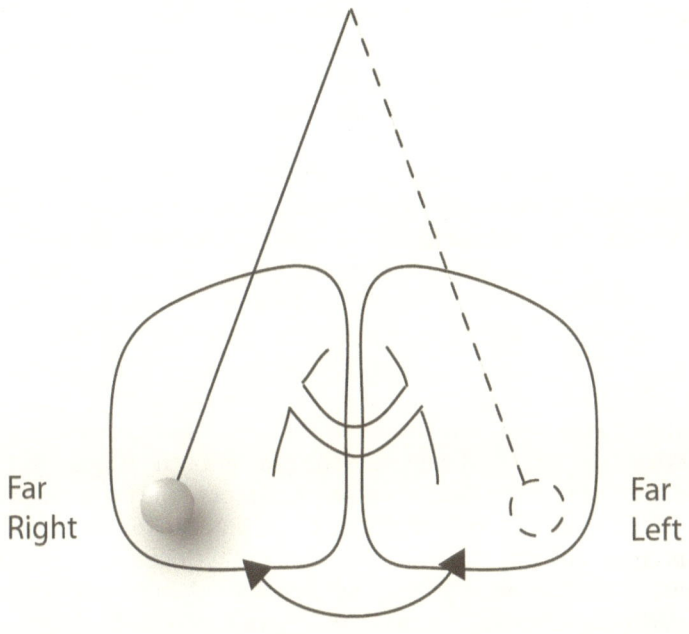

Far
Right

Far
Left

Pendulum Bipolar Oscillation

Galileo experimented with pendulums and made several hypotheses about its behavior, including his discovery that the period or frequency of the pendulum cycle is almost the same, regardless of its amplitude or arc.

his professors, he was able to break the *Law of Diminishing Returns* and build a new growth cycle for himself. Newton, along with Leibniz of Germany, would discover Calculus as a necessary tool in determining problems such as the area under the curve of a growth cycle, or even the maximum combination-point for the same curve to produce optimal output. And in the field of optics, he "discovered" that white light is not a homogenous entity as his professors said. He projected a beam of sunlight through a glass prism, and he noticed that it split into an elliptical spectrum of colors—red, orange, yellow, green, blue and violet. (We now know, of course, that rainbows are caused by water droplets in the air, after it rains, reminding us, that unless we accept and harmonize the rainbow of human races and personalities, we will all perish. And in the temple of Heliopolis in Ancient Egypt, patients were treated in color-therapy rooms designed to split the sun's rays into the spectrum of colors.) Naturally, Newton also experimented with gravity, which led him to a simple formula for the *Force*—the power that may bring everyone finally together.

Because of his discoveries, he was promoted to Lucasian professor in Cambridge. But he actually would have laid aside his new knowledge, if not for his friend Edmund Halley (known for his re-discovery of a comet), who persuaded him to document his findings. So, in 1687, the *Philosophiae Naturalis Principia Mathematica* (The Mathematical Principles of Natural Philosophy), or *Principia*, was published.

But in 1693, he suffered through a *nervous breakdown*. In spite of his mental illness, he was able to eventually seek a government position in London, and in 1699, he became the Master of the Royal Mint. He would then go on to be elected as the President of the Royal Society in 1703, and then knighted in 1708 by Queen Anne. Ultimately, in 2003, he was chosen by popular vote, through a BBC documentary and the internet, as the "Greatest Briton," who ever lived, followed, of course, by Sir Winston Churchill. Indeed, the *Force* was with him!

Newton's findings would later be used by Einstein to solve the deficiencies with Newton's Law, which uses Euclidean geometry. Eventually, in 1915, Albert Einstein submitted his final version on *The Field Equations of Gravitation* or better-known as the *General Theory of Relativity*, which created a "law of gravity" that is applicable all over the universe. Now, the true nature of the *Force* is known!

Clearly, to make astrology more accurate, astrologers must use tensor calculus and differential geometry, since the universe relies on non-Euclidean geometry. Astrologers also peg their personality spectrum to seasonal indicators such the summer solstice, but the earth at perihelion changes by one full day every 58 years, which varies personality links over the centuries. They need to make their personality charts dynamic. This looks like a job for IMA (Institute of Mathematics and its Applications) in Minnesota, USA, which may produce an algorithm for it, if they haven't produced one already, for medical purposes, to help people understand themselves.

Actually, Traditional Chinese Medicine (TCM) already uses the 5-element theory to prescribe herbs and other treatments for people with ailments. It is also interesting to observe that while half of top corporate executives believe in "astrology's" predictive ability, only one-third of the general public accept it as true. This is probably due to an executive's ability to understand the cycles and to manipulate it. However, two-thirds of the populace believes in its ability to influence personality.

Regardless of what you believe in, there is no doubt that God will always play a role in your life. Indeed, in this case, God would be the *Tao*—Gravity, Space and Time—as the ancient Taoists intuitively proclaimed it. It is what created and shaped the universe and yourself, like a Chinese Master Potter, molding his clay with a spinning wheel, just as he did 7,000 years ago. Gravity, as Einstein explained it, is the effect of the curvature made by a mass, such as the sun, in space-time. If you place a baseball for instance on a pillow, it produces a curve. And if you place a marble on the sloping curve of the pillow, it will fall toward the baseball. The pillow here is space-time and the falling marble is the effect of gravity. But the planets orbiting the sun, "fall" perpetually, circling the sun like a whirlpool, with the force of gravity not strong enough to suck the planets toward the sun. This is how the *Tao* created the solar system, by swirling pools of gas and dust together and producing several massive spheres, or planets.

Then as the *Tao* kept spinning its wheel, it produced various flora and fauna on earth, together with the human races and its legion of personalities. This is why even if certain personality types or people are exterminated, as eugenicists would wish, through ethnic cleansing, neuroleptic drugs, anti-cycle wars, artificial viruses, gene therapy and the like, they will, most likely, "grow" back, in due time, even if it takes a very long time, since the *Tao* will simply recreate them. Actually, it is possible for certain races to become extinct, just as the British exterminated many Australian and North American aboriginal tribes, when they grabbed their land, but among the races still left, their "personalities" will be recreated by the *Tao*. And even if the extermination process is continuously done thoroughly, and only one race is left, the personalities will still be recreated within that race.

This is why the spectrum of personalities will always be around, until the earth reaches its end.

Indeed, this is probably why the Buddhists say that there is no such thing as "good" or "evil", "normal" or "abnormal," "superior" or "inferior," "sparing" or "prodigal," but only relativity and two-dimensional bell-curved thinking that makes it so. For instance, some people think that males are "evil," since there are more males in prison; and females are "good," since they care for their young more than the male. But others counter argue that females are emotional, weak and indecisive, while males are strong thinkers who have a variety of functional skills. Indeed, this is why some cultures prefer males over females, and they call the man as "good" and the woman as "evil," just as symbolic myths like Eve's forbidden fruit and Pandora's Box highlight. And if you go with the bell-curve definition of what is "normal" based on human population, the Chinese will be considered as "normal," since they are the most populous, and everyone else is "abnormal." But I'm sure that George W. Bush would disagree with that. Moreover, Mensa people think that having a high IQ is "superior," but there are many people who have relatively lower IQs and higher EQs who are more functional than the Mensa crowd. Indeed, when Carnegie Mellon University's *Tartan* reviewed my second book, it revealed their failure to understand Lewis Carroll's wag-the-dog definition of insanity, indicating their lack of common sense and highlighting the stereotype of the "intelligent" student who is dysfunctional in the real world. In the end, the same is true for other opposing concepts, and what is "this" and "that" is a matter of *relativity* or position. More appropriate terms would be "right," "center" and "left." This is why Leonardo de Vinci, in painting his masterpiece, *The Last Supper*, modeled the face of Jesus and Judas from one person, because they are actually the "same."

So, does that mean that if George W. Bush, Osama Bin Laden and Saddam Hussein are all not necessarily "evil," can anyone just do anything that he or she wants, just like female Praying Mantises and Black Widow spiders eating their male mates after copulating, or just like business partners who kill each other after a bad deal? Certainly not! That's the reason why the Middle Way was invented by the Ancient Asians, to stop the ball from spinning perpetually, and to stop the viscous cycle of suffering. It means people need to harmonize themselves and their relationships with others, so that as much as possible, everything is balanced, so that eventually, all people "win." This is the objective of the *Noble Eightfold Path*, which speeds up convergence and produces relative peace (absolute peace can never

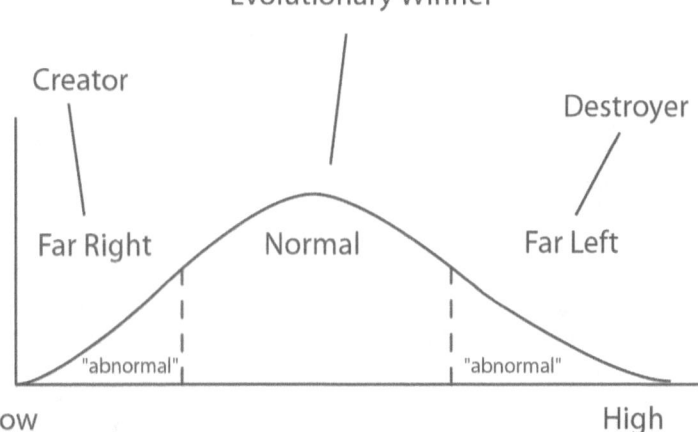

Frequency Distribution Curve

Entities at the extremes are almost the "same" since their results are "equivalent." For instance, someone whose IQ is too high may end up in the same dirty streets as someone who has a low IQ. Or a black man who is worried about poverty in Africa shares the "same" feeling as a white man in the USA who is worried about losing his assets. Hence, the Chinese and Indian-Malays evolved to be the most populous and will likely be the preserved race of the future., since the two extremes tend to move toward extinction, unless conscious intervention takes place.

be achieved just as a hyperbolic curve can approach its asymptote but never touch it).

Hence, if relative peace can be achieved within the individual by balancing the five elements, then if its equivalent—the five faces of society—is also balanced in terms of social, security, economic, spiritual and political policy, then relative peace can be achieved within the family, city, state, country, region and world. This is what Master Kung Fu tried to achieve, as he unified and pacified China. And he states this learning theory, of individual change first, moving outward, as the *Eight Steps of Learning*.

This is what I tried to do when I documented my progress on self-learning in my three books. It is interesting to note that during this millennium (2000 to 3000 C.E.) the planet Mars will be setting new records in recorded history. It will be closer for *eight* times during this millennium, and therefore more influential through its *Force* to the earth since approximately 60,000 years ago. The first record-breaker occurred on 8-27-2003 at around 1800h, GMT +8, my time zone in South East Asia. It was the eve of my mom's birthday, on the 28th, and Mars was about six times larger and 24 times brighter than usual. It was also called Mars at Perihelion and Opposition in 2003. And according to Myles Standish, a NASA-JPL planetary-orbit expert, the next record-breaker will occur in 2287. And the closest of all eight comes in 2729, if the earth is still alive at that time. Hopefully, though, humans will learn to melt their hot Martian-selves so that people can go on living throughout the future. And this is what I hope to accomplish through my three books.

In my first book, *Little Voices: A True Paranoid Schizophrenic Adventure* (ISBN: 0-7388-2327-9), I narrated the story of my trip to the Southern Californian "Wonderland," where I began to hear five voices after I had a breakdown there. I discussed how Western doctors and Western-trained physicians could not help me. I found it very difficult to live during those first few years. But after several years, I discovered that Asian philosophy actually had a reasonable analytic explanation to what my voices said, which Western doctors could not interpret correctly.

I eventually realized that my five voices directly corresponded to the five elements or the five needs or the five faces of society within the individual mind. Everyone actually has five similar voices, but it is usually not that audible, and the number of voices would vary, more or less, depending on his or her brain chemistry. But during a schizophrenic episode, you start to hear yourself—thinking that it is an external voice—as a result of the mind's reflex attempt toward self-protection and balance, similar to the reflex of sneezing or coughing during a viral or bacterial attack. In my first episode, the voices actually sounded like they were coming from a 3-D 5-speaker surround-sound system, since it was directional, seeming to come from distant objects. And although they did not have names, I gave them descriptive names. Three of them spoke in Tagalog or Filipino, and the other two spoke in English with an American accent. They all spoke like teenagers. I called the first voice as *Inoe*, since she was usually silent, saying only

"I know," in agreement with the other voices. The second is *Feelsoree*, since he was usually pathetic and said, "I feel sorry for this guy." The third is *Oobola*, since she usually said, "*Ooooohhh, bola!*" with a varying pitch, meaning, "Awww, c'mon. That's just spin!" or "Awww, c'mon. You're lying!" The fourth is *Tanga*, which is essentially what he said. It means, "Stupid!" And the last one is Ulol, which means "Mad!" or "Crazy!"

I became crazy as a result of interrogating the voices and sometimes following them, answering their call for me to be a "prophet." Fortunately, I was relieved that many ancients, such as Abraham, Moses, Socrates, Jesus and many scientists, have experienced this phenomenon in the past, and that I was not alone. And after a few years, I realized that our minds are like political forums. Just as MPs, congressmen and senators debate issues, policies and bills before it is passed, your voices or mind will deliberate a decision or any action, before you do anything. Usually though, it is done silently, unlike in a schizophrenic episode where the voices become very audible.

My voices eventually disappeared, one by one, leaving only one voice behind, Oobola. But instead of saying negative words, she would say positive statements, usually attaching "They say…" or "*raw*" in her sentences, such as "*They say* you are not crazy" or more literally, "Hindi ka *raw* sira ulo." Or "*They say* the doctors are stupid. They are talking about the doctors." At first, I thought my voices simply coincided with my ego (*Inoe*), superego (*Feelsoree*), id3 (*Oobola*), id2 (*Tanga*) and id1 (*Ulol*) or my five elements. When I consulted Master Allen Tsai's free Five-Element calculator in www.chinesefortunecalendar.com just recently, I found out that my ego (which he calls "you") is 120, superego ("mom") is 60, id3 ("kids") is 23, id2 ("money") is 34, and id1 ("job") is 97. It is very interesting to note that my lowest level is id3 or Oobola, which Tsai calls as my "lucky" dragon ball, since increasing its level or pursuing it will bring balance to my five elements. It is also interesting to note that *Oobola*, my lowest level, was the loudest and the most vocal of my voices, while *Inoe*, the highest level was soft-spoken and the most silent. Even the position of my American and Filipino voices seems to be in opposite places, since the Americans were cruel. So I concluded that our bodies try to automatically balance itself, and it tries to pursue activities as well, that balances your five elements. Indeed, being fluent in more than one language opened a door in understanding the five element spectrum in the mind.

But recently, I discovered the science of brainwaves. Your brain emits various waves in different frequencies when you think or even when you are sleeping. In fact, scientists have already developed computer

Five-Speaker 3-D Surround Sound System

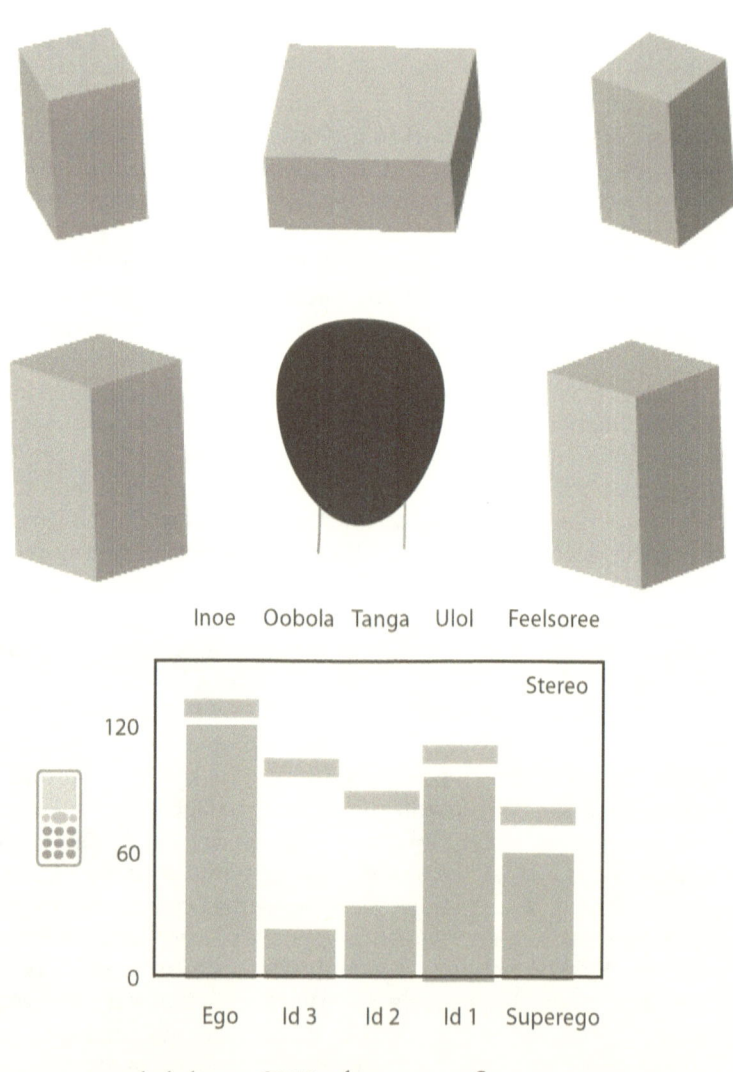

Inoe   Oobola   Tanga   Ulol   Feelsoree

Stereo

Ego   Id 3   Id 2   Id 1   Superego

6-Way Teleconferencing with Bluetooth Headset

games that can be controlled by thought, and they are currently working on other applications such its use in aiding paralyzed people to be able to communicate with others without speaking. They will type in a computer with their thoughts. So, if people transmit brainwaves, can they receive it?

Engineers build large V-shaped pyramid-tower antennas or parabolic-circular dishes to receive radio waves from transmitters. If you look at your head and your body, it is shaped like an inverted pyramid with the centers of gravity being the vicinity of the hippocampus in the brain and the heart, respectively. Theoretically brainwave reception seems possible, but on a practical basis, some people have actually demonstrated it. Yuri Geller, for instance, appeared a few weeks ago in CNN and confirmed brainwave transmission and reception, just as he has done in other television programs and even with actual physicists in different universities. The newscaster, Becky, drew a picture of a circle with V-shaped "hands" inside it, and a number 3 on the right side of the interior of the circle. In other words, she drew a clock, and she didn't show it to anyone. CNN, of course, swears to the validity of this. Then Yuri asked her to think about the drawing, and he drew a circle with the two V-shaped hands facing the opposite direction. In other words, it looked like a mirror-image of the original, but without the number 3. Yuri then explained that the reason why he didn't get the number 3 was because Becky didn't think about it. In other words, he cannot really "see" the picture. He can only read it from your mind, if you are thinking about it *consciously*. (It is, however, not recommended to appear publicly like Yuri if you can do this, since the consequences are not favorable.)

Finally, when I thought about my first theory on my five voices—that it is my ego, superego and three ids—and when I also thought about brainwave theory, I concluded that mostly likely, my five voices are a combination of both theories.

In my first book, I also illustrate some graphical ideas or "visions" that I thought of during my first episode in 1994. After several years, I discovered that they were actually *mandalas* that are useful for Visualization Meditation (focusing on the image), choosing an image from the selection, depending on who you are.

These *mandalas* show how the *Tao*, like a Master Sculptor, chiseled you into who you are, and intuitively teaching you the *Way*. The human body's V-shaped anatomy and spherical contours show the influence of Gravity in shaping humans, just as it shaped the planets with the "Compass" and the "Circle." People's "artificial" attraction to a pair

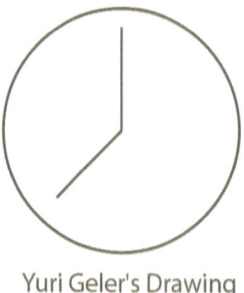

Yuri Geler's Drawing

Becky Anderson's Drawing

Time to Sync

Brainwave Live Demo authenticated by CNN's Richard Quest and its producers while I thought about the compass and the circle. Becky drew a clock, but Yuri drew a mirror image of it.

of spheres, whether it be the eyes, breasts, buttocks, or even one's brain—which is actually two hemispheres, a left and a right one pieced together like a woman's buttocks—shows how the *Tao* inclines you toward convergence, making you fall in love with the Middle Way, the harmony or union of right and left, good and evil, or female and male.

My first book also gives *mantras* (meaningful words)—three explicit on the light-sided faces and some implied on the dark side of Pandora's Box. These "mantras" came to me during my third "episode" in Mindoro Island, the heart of the Philippines, when I did not realize that they were *mantras* useful in Concentrative Meditation (focusing on the syllabic meaningful words while breathing in and out.) Together, these *mantras* and *mandalas* become useful tools in Loving-kindness Meditation.

In my second book, *Illuminati: Healing and Developing the Mind* (ISBN: 0-595-65472-X), I discuss the Middle Way for the individual as a "cure" for mental illness, balancing out your brain chemicals or five elements. I also lay out the beliefs of the extreme left to the extreme right, in a table of ten theories, which I call the *General Theory of Everything*, since it is from the *Trinitarian Spheres* of the Middle Way were everything emanates from. Then I explain the *School or Learning*

Gamma 40 Hz	Anxiolytic or anti-anxiety, learning, information processing 40 - 60 Hz stimulates release of brain chemical beta-endorphins	
Beta 14-40 Hz	Alert, Awake, focused mind	
Alpha 8-14 Hz	Deeply relaxed, creative, meditative state 9.5 Hz frequency linked to earth's magnetic field 10 Hz brain chemical seratonin released 10.5 Hz healing of body and mind	
Theta 4-8 Hz	Drowsy, deep meditation, intuitive 6 Hz stimulates memory	
Delta 0.5 to 3.5 Hz	Sleep, healing and body regeneration, unconscious 3.5 to 2.5 Hz insomnia 1.5 Hz aids chronic fatigue	

Theory, where the judge says that everyone is just the same, as guilty as everyone else, and where "angels" and "demons" become "identical." Consequently, people can learn how to get along with each other through the Middle Way, where conflicting parties are encouraged to be both "winners" as much as possible by *balancing* out. It also leads to the understanding that *every person* has abnormalities and imbalanced brain chemistry. Hence, everyone needs to know how to manage it, by living a *moderate* life, since you can't just do anything that you want and tell the victim, "My genes made me do it!" No court of law will accept that defense.

In the "hidden" second half or dark side of my second book, I tested and confirmed one of the implied *mantras* whose algorithm can be used for the light side, but dangerously also for the dark side. I was able to predict the September 11 Tragedy, ten months ahead, posting proof in the internet, and predicting also the "split world" which would come as a result of it. I did this by posting a picture of my second "vision" in my website last 11-11-2000, as a schizophrenic experiment. To my surprise, it proved correct! In this sense, I capture one of the "tricksters" and insert it back into "Pandora's Box" from the first book, authenticating a blunted version of *Theory X=Y+Z*. Of course, this *modus operandi* won't be effective anymore, as it was in World War II. As the world says, "Fool me once, shame on you. Fool me twice, shame on me!"

Clearly, the Bush strategists made a miscalculation, and it will cost America very heftily. In early 2003, Bush, with his fiery rhetoric, threatened to divide Europe for condemning America's psychopathic behavior. But later in that year, on September 7 (GMT-5), Bush admitted that he needed their help. He was falling, just like Tony Blair, and he couldn't get up. (One of his puppets, Gloria Macapagal-Arroyo also suffered under a military mutiny in the Philippines.) A day later, on 9-9 (GMT+8), North Korea celebrated its 55th anniversary with a military parade in Pyongyang, scoffing at Bush's rhetoric. Interestingly, the media was able to capture a scene where a bus-load of North Korean cheerleaders were all upset about a banner with Kim Jong Il's picture on it. It was left out in the rain and was hung crooked or tilted. They thought it was sacrilegious negligence to leave their leader's banner imperfectly hung, so they tried to fix it. And after they fixed it, they were happy again. Clock King would have suggested a nano-precision ruler to hang their leader's banner. And it makes you wonder who is worse? The Ventriloquist and Scarface? Or Clock King?

So in this third book, I will be discussing the Convergence Theory as it manages the five sectors of society, leading to a more balanced guide to policymaking, not only for individuals, but also for organizations and politicians. Social, security, economic, spiritual and political development is usually thought of separately. But it is actually all inter-related. And to consider only one sector, while ignoring the rest, leads to failure. In organizational development for instance, even Western change-management "experts" or consultants only have a one-third success rate at the moment. The reason is simple. They usually use techniques from the id, such as force, intimidation and threats, leaving out the spiritual dimension (superego), forgetting to listen to the five voices or sectors, failing to analyze their needs, and unable to harmonize it (ego). In other words, two-thirds of the "big picture" is left out.

It is also interesting to see how the West became successful by using Eastern ideas, ever since the ancient Greeks and Romans, up to the time of Freemasons such as Ben Franklin and George Washington, who established their empire by using the dark side of Eastern philosophy. On the other hand, it is tragic to see how parts of the East became failures by using Western Marxism, Christianity, Capitalism and the American presidential system, much of which was forced down their throats through colonization, or established because of security needs. You do know, however, that opposites attract, which is why they try each other's systems out, providing checks and balance to the world. If the East didn't check on the West and vice-versa, you won't arrive at the correct conclusion.

When it comes to holistic well-being, however, Lewis Carroll is right. If I didn't escape from the "American Wonderland," toward the ASEAN region, I probably would not have healed myself. Charles Dickens is also right when he wrote *American Notes*. Although it was badly received by the Americans because of its direct assault on them, Dickens' *A Christmas Carol* had a better reception, since this time, he portrayed America as the world's Mr. Scrooge, overlooked by most Americans. Today, of course, these books are not potent enough, so modern writers such as Bret Easton Ellis write new books such as *American Psycho*; this time, graphically depicting America as a serial killer. Naturally, the American media critics attacked Ellis for his blatant portrayal of the American psyche. But a movie adaptation was created recently for that book, starring Christian Bale. And interestingly, Bale is in the shortlist for the role of Batman in an upcoming movie for that series. Perhaps Americans are now becoming aware of their psychopathic selves. Let us not forget, however, that there is still a minority of Americans, about one-third, who know that their government needs to be re-assembled, without its destructive potential. But you do have to manage your expectations, since there is a limit to how much and how fast people can change. So instead of trying to wish for the stars, we shall try to see how the *Tao* molds people in all levels of society, so that we can alleviate fear, anger, sadness and other types of emotions that are caused by poorly-managed *expectations*.

In Chapter 2 of this book, I will summarize the main points of my first two books, to converge the ideas presented in it. My first book is circular by it introduces the "Compass." On the other hand, my second book is a compass, but it introduces the "Circle." Its complementary attributes will be highlighted in this chapter to see how it fits together, forming a bigger picture. Here, we also try to listen to the five voices so that we can try to harmonize and address their needs. Listening doesn't mean literally following one or two voices. It means that you should take note of their needs and try to figure out a creative solution to balance the five needs.

In Chapter 3, I will discuss *social policy*. I will be questioning Roman law and its counterpart in ancient Asia-Pacific social customs, which seems to be more harmonious than practices such as the labeling of offspring as "illegitimate" (unmarried parents), "legitimate," and "legitimated," while being granted tax or inheritance benefits for having the right labels. I will also be investigating marriage customs around the world. Tibetan teenagers, like other ancient Asia-Pacific teens, are free to sleep with each other without the blessing of marriage, but are also free to marry if that is their wish. In fact, in the ancient Asia-Pacific, marriage was only initiated on the children of conflicting tribes and kingdoms, to bring peace. Of course, some Asian countries are now Christian and Muslim, which bans pre-marital sex. And in today's overpopulated world, some measure of control is needed to produce harmony. What is the right balance?

Naturally, in Tibet, the people eat a lot of peas, which may be a natural form of birth control, since their birth rate is very minimal. But more potent methods of birth control are needed in countries such as India and China. In the ancient past, men had to produce many offspring with multiple partners since many of their children were expected to die prematurely through natural disasters. But today, nature's way of population control such as wars, famine, disease, typhoons, hurricanes, earthquakes, and other natural disasters are now less effective in managing the population due to human intervention. So in the future, you will probably see large populations dying at the same time due to the obvious elongated load. Those who are left behind will simply repopulate the earth.

Indeed, in Buddhism, dying is not really considered as a "bad" thing. There is a Buddhist story called the *Soliloquy of the Frogs*. And in this tale, an elder frog explains to the young ones that their beautiful surroundings are there for frogs—heaven, earth, water, air, and even the insects that they eat. That made the young frogs jubilant, but all of a sudden, a snake jumped out of the bushes and ate one of them. So a baby frog asked if the snake is there for them too, and the elder replied, "If there weren't any snakes to eat some of us, then we would over-multiply, and then there wouldn't be room for all of us." Certainly, yin-yang in nature will increase "evil" if there is too much "good," and vice-versa. That's why, even if George W. Bush seems to be a cobra, it's just a natural part of the universe. And if you manage to escape from his fangs, consider yourself lucky! (Of course, snake hunters who catch cobras for their medicinal venom are also part of nature.) And even Roy Horn of *Siegfried & Roy*—the German American magicians from Las Vegas—did not condemn his albino tiger when it mauled him recently in their last performance at the Mirage hotel and casino in 2003. He is now paralyzed in the hospital, after instructing his staff not to hurt the tiger. The albino tiger, just like Hitler, is part of nature. That's really the way they are and that's how nature balances itself. (Of course, other types of people who would retaliate in such a situation are also part of nature and that's how they bring balance to the *Force* or the *Tao*.)

How about lesbians and gays? Gays in the West receive a significant amount of prejudice, as compared to the East, where they are thought of as humorously weird but do not receive as much abuse as they do in the West. Should they be allowed to marry? How about adoption? We do know the story of the mouse which asked for a slice of cheese and was not contented, so he asked for a glass of milk, and so on. Can unorthodox marriage lead to the reduction of disease epidemics, Roman orgies, Sodom and Gomorrah, the Temple of *Kama Sutra*, and human trafficking for sexual exploitation? We do know that the acidic nature of imbalanced people is neutralized through the proper bonds. It certainly reduces crime, but what is the right balance?

We also know that contrary to what many non-whites say, Westerners have actually contributed in reducing many savage non-Western practices, such as the use of large lip plates on African women and neck rings on Paduan women (Imagine cutting your lower lip and inserting a saucer in it! Or think about inserting rings around your neck to extend it like a giraffe!) And if Western women did not criticize the ancient Chinese practice of feet binding, the Chinese government probably would not have banned it. And if the British did not colonize India, women there probably would be much worse off than they are

Just as I predicted last year in the 10th chapter of my 2nd book, Bush's limo, with a US Flag on its right headlight and a presidential flag on its left, passes by the street along my residence on Saturday, October 18, 2003 in Metro Manila, Philippines.

A paranoid convoy of white and black vans carrying the U.S. Secret Service trail Bush's limousine, which passes by my place as its female driver speeds past a cop blocking my view.

President George W. Bush visits the phalic-masonic Rizal obelisk after his arrival. Bush lays down a wreath at the stupa-shaped monument, remembering the author of *Noli Me Tangere* and *El Filibusterismo*. He then proceeds to Malacanang Palace, where the First Lady, Laura Bush, tells a story about various animals to a class of school children. The children reciprocated by presenting a skit called the *Story of the Bamboo*--a Philippine folk tale similar to the *Soliloquy of the Frogs*. The story narrates a bamboo that complains to a farmer who keeps cutting it down. But the farmer replies that if he doesn't cut the bamboo, remove its branches, *split it into two* and carve out its guts, it wouldn't be useful. So the bamboo conceded and was turned into pipes that watered a rice field and produced a bountiful harvest. The school children then presented other bamboo products to Laura, such as flutes and stupa-shaped fish traps. I guess that's why the Californians split me in half. And maybe that's why the Muslims blow up Americans. But thank God that George arrived safely back home in the U.S.

Sir Elton John's Circle of Life

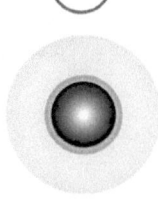

Sun kills bacteria, viruses and the like

Plants consume sunlight for photosynthesis

Bacteria, fungi, viruses and parasites feed on humans

Primary consumers such as rabbits feed on plants

Humans feed on primary and secondary consumers, plus plants and sunlight

Secondary consumers such as coyotes feed on primary consumers

Unlike what some people think, the Food Chain does not have a "top" or "bottom." It is circular. Humans, for instance, who are thought to be at the "top" of the Food Chain is consumed by bacteria, which is thought to be in the "bottom" of the chain.

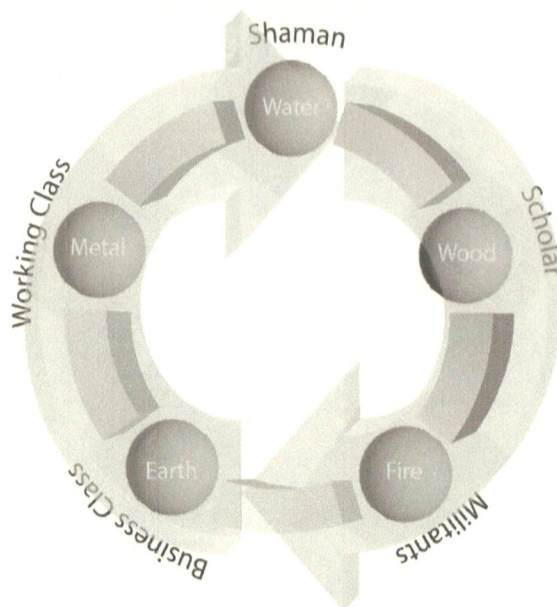

Ko or Regulating Cycle

Different types of humans also "eat" each other, but the Ko Cycle shows how they ideally help each other. This affinity paradigm shows how "Water" nourishes "Wood" or "Trees." "Wood" builds "Fire." "Fire" produces "Earth." "Earth" forms "Metal" or "Minerals." "Metal" then holds "Water." This is how they theoretically assist each other. In an organization, for instance, the O.D. Trainer or Shaman advices Top Management or the Scholar.

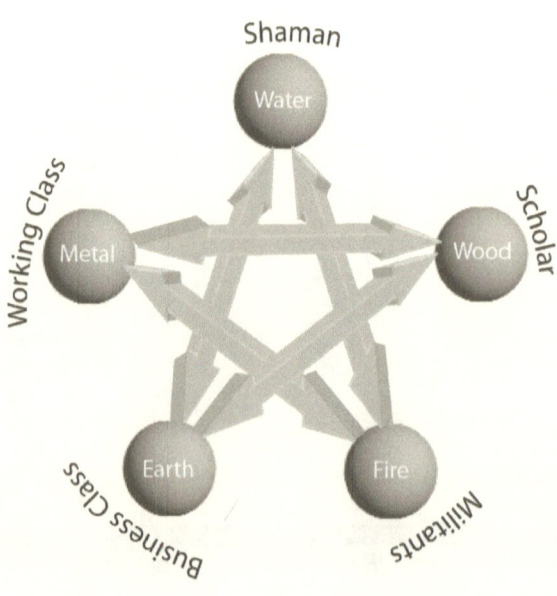

Enmity Relationship

Thousands of years ago, the ancient Taoists developed Conflict Management principles, and they concluded that trying to make everyone completely harmonous with each other is impossible. But they developed this Enmity Paradigm to understand how various types of people fight each other. In doing so, they were able to minimize or manage conflict. For instance, "Earth" or Sales and Advertising might be in constant conflict with the O.D. Trainer, since he warns them against untruthful advertising and sales tactics which might place the organization in trouble with the government or consumer today. (Just imagine the carvings in the Temple of *Kama Sutra* and think that you are a woman! It certainly might be worse than being enclosed in a glass cage in the Netherlands!) Certainly, if the Spaniards did not colonize the Visayans of the Philippines, they would probably continue to sport their tattoos. And in these cases, Westerners did some good for society, especially for women who no longer need lip plates or bound feet to be able to find husbands. (Westerners though continue to sport the unsightly Asian practice of wearing make-up.)

In the end, we know that sex is a physiological need, but marriage is a superego need. Tax and inheritance benefits are security needs, and the need to be accepted by society is a social and belongingness need. Your policy then will depend on what you really need.

In Chapter 4, I will discuss *security policy*. The world knows, of course, that the Anglo-Saxon Celtic Empire is currently the big bully of the world. He is fully armed and there is no counterbalance to check on him. His no-negotiation win-lose policy irritates diplomats all over the globe. We do know that his militant nature is natural to him and that as long as this imbalance is not neutralized, terrorism will continue to thrive and contaminate everyone else in the world, even the bystanders. Should the voice of the militants be suppressed with the WMD called Vx Nerve Gas, just like psychiatric neuroleptics which are made of the same ingredients, or should their needs be understood, addressed and neutralized? We do know that North Korea was compliant with the International Atomic Energy Agency (IAEA) until Kim Jong Il tried to sell Korean clothes to Bush and the American market. Bush hates "pygmy" products so he cut off Kim's fuel supply. That provoked Kim, leading him to kick out the IAEA in his country, and then he started to produce nuclear bombs to ensure North Korea's security. The world, naturally, knows that the U.S. really needs a lot of therapy, but in the end they were willing to set six-way talks with North Korea, after Bush's initial no-talk policy. Kim agreed to dismantle his nuclear program if the U.S. signs a non-aggression pact. Fair enough. But the U.S. didn't want to sign anything, and it really became very clear that what they really want is to wipe out the Clock Kings. Even a high school student who studies Algebra (*Al Jihbra*) can easily understand this. The Arab Asians invented this branch of mathematics as a way to provide balance and restoration, which is what its name means. But Bush's math needs a lot of work:

Current:
$$WMD_{USA} = Deficiency + WMD_{World} + WMD_{Korea}$$
Bush:
$$WMD_{USA} + Death\ Star_{MDS} = Deficiency + WMD_{World} + WMD_{Korea} - WMD_{Korea}$$

Any decent high school math student can easily see that Bush-type Americans don't understand the *Additive Property* of Algebra—you can't add on one side and subtract on the other side. Indeed, for the world, 1+1=2, but for Bush-like far-right-far-left bi-polar Americans, 1+1=911.

How about correctional facilities? The U.S. currently has one of the highest, if not the highest rates of incarceration. You just need to tune in to the *MTV channel* or the *Cartoon Network* or comedy shows or to American magazines such as *Time* to get a glimpse of all the mental disorders of many Americans. Of course, at least they have a channel to unleash their emotions, unlike other countries that do not have that freedom. On the other hand, take a trip to Basilan Island, in the ASEAN, and you will see that their jail house is empty. Take another trip to a correctional facility in Palawan Island, which looks more like a self-sufficient beach resort. They farm their own food and carve their own crafts for sale. Their families can be with them. And when they are released, they end up becoming saner than white-collar New Yorkers running in the Rat Race, but doesn't that make the criminal better off than the average working man? Should this be propagated or should the death penalty just be implemented to save tax money? What is the right balance? We do know that many innocent people end up behind bars and many criminals end up in board rooms and government offices, thanks to Two-Face.

It is also interesting to note that about 2,500 years ago, Sun Tzu wrote the greatest book on security policy and strategy—the *Art of War*. Even American businessmen and multinationals greatly revere this book for its applicability to business enterprise. Master Sun actually uses Taoism, the duality of yin-yang, and the principles of the five elements to systematize military strategy. He states that warfare must be compared and calculated by using five perspectives; namely, (1) the *Tao* or establishing a moral cause, (2) "Heaven" or the forces of nature, (3) "Earth" or terrain and geography (4) Command or leadership and (5) Law or discipline. Then he uses these five principles in various combinations to explain strategy. As Sun says in Chapter 5, "There are only five musical notes (in Asian music), yet one could never hear all their possible melodies. There are only five colors, yet one could never see all of their possible shades. There are only five tastes, yet one could never try all of their possible blends. Frontal and surprise confrontations are the only two kinds of force for waging war, yet their possible combinations are limitless. Their mutual transformation is like tracing the line of a *circle*—there is no endpoint." He also mentions in Chapter 6 that "Deploying forces is like the transformations among the five elements—metal, wood, water, fire and earth—alternately arising and giving way, not concerned with which one is at a temporary advantage." He also states five personal traits that are dangerous in a commander:

(1) He who is intent on dying can be murdered
(2) He who is intent on living can be captured
(3) He who is quick to anger can be insulted
(4) He who is self-conscious can be humiliated
(5) He who is compassionate can be troubled

And to politicians, Sun Tzu says, "you must act according to what benefits the country. If it is not beneficial to the country, halt activity; for anger can become happiness, and indignation can become joy. But once the country is lost, there is no way to bring it back; and once a life is lost, there is no way to revive it. Therefore, an intelligent sovereign will weigh heavily his choice of engaging in war, and an outstanding general will use his forces with caution."

Another Asian book about security and strategy is Miyamoto Musahi's *Go Rin No Sho* or *Book of Five Rings* (1645 C.E.) which explains *Ni Ten Ichi Ryu* or the Way of Strategy. It explains the strategy of the Five Faces of Society and he compares it with the martial arts. He analyzes strategy in terms of Void, Fire, Earth, Water and Air. He also mentions these as the way of the "farmers," "warriors," "merchants," and the "artisan" or "carpenter" (the two-faced shaman-scholar), respectively. Just like Jesus the carpenter or a mason, or "like a trooper, the carpenter sharpens his own tools. He carries his equipment in his tool box, and works under the direction of his foreman. He makes columns and girders with an axe, shapes floorboards and shelves with a plane, cuts fine openwork and carvings accurately, giving as excellent a finish as his skill will allow. This is the *craft* of the carpenters…The attainment of the carpenter is that his work is not warped, that the joints are not misaligned, and that the work is truly planed so that it meets well and is not merely finished in sections. This is essential." It is also interesting to note that the shape of the Pentagon building—home to the U.S. Department of Defense—is modeled like the Five Rings.

In the end, trying to achieve the right balance for security is tricky, but there are certainly ways to minimize the insecurity of people around the world, addressing their need for violence and justice at the same time.

In Chapter 5, I will discuss *economic policy*. The world knows that 50 % of the world's resources in consumed by 5% of the world's population, the Americans. In 2003, the U.S., along with Canada, experienced its worst black out in its history. For a while, New Yorkers were in the dark. London, England, and even Rome also had similar blackouts. It seems that their infrastructure is taking its toll already. On the other hand, Singapore discourages the use of cars by taxing it heavily to conserve energy and space. Developing a clean, fast and efficient public transportation system is their priority. What is the right balance between the two?

How about anti-cycle policies? Do they make things worst or better? Can the business cycle be stabilized? Can monetary and interest-rate policies simply do the trick? Or must the entire system be overhauled? We do know that the economics of Karl Marx doesn't work, but neither does Capitalism. Governments spend more now on military development than on human or organizational development. Economic sabotage on third-world countries is often disguised by the West as "economic aid." Farm subsidies and genetically-modified agricultural products, for instance, make U.S. food very cheap in the world market, killing third-world farmers. By distributing free U.S. food in developing regions, such as Africa, they destroy their livelihoods further. How about the WTO, IMF and the World Bank? Can they be moderated? How about the American Big Boy's Investment Club? Can regional security laws be implemented to prevent them from staging another Asian Currency Crisis or another U.S. stock-market bubble? Whatever also happened to anti-monopoly laws? Why were there so many mega-mergers created in the last few years? Can small businesses survive with giants around?

It is also interesting to note that, in 2003, the California State Assembly approved a soda ban in elementary and middle schools. Can the world follow suit and block harmful American products from their shelves. Not very many people know that soft-drinks are acidic enough to melt a carpenter's nail in about 4 days. Its active ingredient is phosphoric acid and its PH is 2.8. And in many U.S. states, the highway patrol carries 2 gallons of Coca-Cola in their vehicle to remove blood stains in case of a car accident. Coke distributors also use it to clean their truck engines. And commercial trucks also label their load of Coca-Cola syrup concentrate with the sign "Hazardous Material." Can that at least be printed on the bottles and cans of soft-drinks if international trade wars are feared? Coke, however, is useful as a cleanser. It is very effective in removing the stains on your toilet bowl's vitreous china. It can also remove the corrosion from your car battery's terminals, or you can use it to loosen a rusted bolt. Why not just sell Coke in a hardware store?

Would it also be possible for everyone to label GMO products? Hormone-stuffed meat and milk, and mercury plus PCB-stuffed fish are creating a new generation of mutants nowadays. We know that the hormones injected into livestock makes them bigger, but how has it trickled down to humans? About one hundred years ago, the average size of ladies shoes for Caucasians was a size 4. Today, it's a size 9. Bigfoot seems to be on the rise! Can lawmakers control the growth of the Abominable American Yeti or the giant white freak? Or will they ignore this mutant trend?

In addition, can the small Mom-and-Pop Inn entice the tourist dollar more than the multinational hotel? We do know that American franchises and business partnerships have helped thousands of entrepreneurs get started easily. American grocery and fast-food franchises are popular throughout the world, and they have improved the standard of service in the industry. American outsourcing entrepreneurs have also employed thousands overseas, not just in the service industry, but also in manufacturing; although many would argue that they created a herd of call-center slaves ("call girls" and "call boys" working in graveyard shifts), back-office slaves and sweat-shop slaves for exploitation. We do know that outsourcing corrects the imbalance in the world and moves capital and income to developing nations. If American entrepreneurs develop easy-access no-scam programs to train and develop the third world in real business, will that help their reputation?

Western marketers and advertisers are also known for their brainwashing and subliminal techniques to fool you into buying something that you don't really need or want. Nowadays, it is easy to find people who have a pile of merchandise that they never really wanted, but somehow they purchased it. The use of subliminal messages in advertising has long been banned, but today, neuroscientists are still researching for the consumer's "buy button." It naturally has something to do with your five basic needs and how your emotions and memory are attached to these *desires*. Multinational branding often plays around with your iditistic need for sex, power and belongingness, but sometimes they also poke your superego through guilt, depending on the left-ness or right-ness of their targeted market; consequently, should lawmakers intervene?

We do know that the Penguin's strongest need is social and belongingness. The tuxedo is just a front. So if businessmen are isolated, will it heal them, or will it further provoke them? In the end, the Western Id knows trade and money instinctively. Thus, developing partnerships with them is an inevitable part of the business process.

In Chapter 6, I will discuss *spiritual policy*. We do know that communist atheism, which overlooks the spiritual dimension, is ineffective in *motivating* people, as communist China has proven. Indeed, a certain level of fantasy is necessary to egg on people to become productive citizens. The spiritual element is just as important as the other five elements in completing people and making them happy and fruitful. So how can the church or the temple be improved or reformed so that people can once again flock to it? And who shall lead it? How should the various churches be organized? Although Asians practice Taoism, Buddhism and Confucianism as philosophies, they used to practice Animism or nature worship too. Should it be revived to produce a greener, healthier environment? Christian ministers for instance, con Malays into thinking that Jesus will be coming again to clean up the world; hence, they say that you can trash the world since it will be wiped out anyway. Malays take it literally now and dump their trash all over the place, unlike in the ancient past when nature was considered divine. Managing pollution and environmental protection is not their priority. Should Christianity then move towards Gnosticism or should it be gradually eliminated altogether? How about Judaism and Islam? How can they evolve?

Or should the public sue religious organizations for defrauding them? Should lawmakers moderately intervene already when it becomes too obvious that certain extreme religions, like fundamentalist Christianity, Catholicism, Islam, Hinduism and even *serious* Buddhism, incite insanity? Or would government-sponsored public grand debates be helpful? A Christian minister was recently executed through a death sentence after he shot an abortion doctor. It's the typical far-right-far-left passive-aggressive bipolar effect. Even during his last days, he had no remorse for what he did. He simply smiled and said that he was on his way to heaven where Jesus, his savior would lovingly accept him. It was also the Christians who supported Bush in his Iraqi-slaughter campaign, since he is a born-again Christian. Muslim clerics have also been pinpointed as the masterminds of recent terrorist attacks. And we all know what's happening in the "Holy" Land, where people kill each other continuously. Should the government at least put warning signs outside the offices of doctors and at the entrance of churches, mosques and temples, just as they do for cigarettes, liquor and movies? Could they show three different opinions—good, bad and neutral—so that the public can make a reasonable choice? Could they do the same thing with music, movies, books, television, bars and the like, to notify the public so that they can make informed choices?

We do know that people need some "opium" every now and then to heal themselves, but too much is clearly hazardous to your health. Indeed, the illusionist David Blaine, who recently survived a 44-day fast as a "magical" stunt, inspired many Britons. He surpassed the 40-day fast of Jesus and set a new record, even if Blaine was crazy to do so.

We also know that several religions are needed to check and balance each other out, providing for the different needs and tastes that people have. Indeed, as Benjamin Franklin, stated in his autobiography, although he is not interested in going to church, he recognizes the fact that churches have some benefit to society, even if many of their beliefs don't make sense. And because of this, he tolerates their business as long as they do no harm. Nowadays, though, their detrimental effects are greater than their positive outflows. Some people don't even realize that Jesus and Hannibal Lecter, who both love to eat human flesh and blood in a ritual of "Holy Communion," belong to the same category in the human spectrum.

In the end, when the ancients thought that they could improve the concepts of Hindu-Buddhism, they invented Christianity; and when they thought that they could correct the Christians when they became dysfunctional, they developed Islam. Then, of course, when Islam got out of hand, Buddhism made a revival. Just when people thought that their lives were improving through "enhanced" spiritual systems, they never realized that they were just going around in a circle. Interestingly, the *Tao* now recognizes all the arrows that are off-target, and its aim is moving closer to the center. It's realizing that what it really needs is guidance, assistance, healing and community, instead of politics. Its challenge will be in determining the right proportions for each remaining religion to produce healthful living, but in the end, the *Tao* automatically moves people gradually to the right balance.

In Chapter 7, I will discuss *political development*. We do know that a Marxist government is ineffective. The five voices are understandably suppressed in a one-party system since listening to them too much will drive you crazy, but failing to understand their needs also leads to the ill health of the state. Even the American presidential two-party system does not give proper representation; and it has too many checks and balances that promote politics instead of minimizing it to move forward. Shaman types, some militants and business class people, for instance, bundle themselves together under the Republicans, somewhat mimicking the brain's right hemisphere, with the right frontal lobe and right temporal lobe as the most active. The other two-and-a-half voices bundle themselves with the Democrats, somewhat like the left brain, using the left frontal lobe and temporal lobe more actively. But it is an inaccurate model of representation since "female" Rights use the Left and Right, while "male" Lefts usually use only the Left.

More distinct parties may be needed. In Europe, for example, we know that the Green party is part of the shaman class since they are environmentally friendly. We know that Tony Blair's Labor Party represents the working class. Could the world follow that system, so that just by hearing the name of their party, you can immediately know what voice they represent? In the Philippines, they use party names such as *Lakas* (Strong) and *Laban* (Fight). It seems to represent the militants, but corresponds more to the working class and even the business class, since the militants don't have any representation at the moment. It's very confusing. Are these labels used to intentionally confuse the public or is it unintentionally fuzzy?

Understandably though, the two-party American system thinks like a robot (True-1, False-0 and nothing in between). The Europeans and Japanese though have some room for fuzzy logic (True-1, False-0 and Maybe). And the Philippines leans more toward fuzzy logic, where even true and false are fuzzy. Their systems seem to reflect their personalities, from left to right. Even the way they drive reflects their personalities, so it all seems to be natural. The Filipinos, for instance, don't recognize the grid lines on the road and wiggle around the lanes unpredictably like particles, whereas the Europeans and Americans drive in between the lines like predictable waves. (In this sense, Westerners are easier to predict than Easterners.) But since the Filipinos currently use the American presidential system, it makes them all the more awkward and ineffective. They, like many other countries, would probably do better if they adopted a more suitable system.

The parliamentary system seems to be the most sound so far, since it seems to be more centered, and in this system, politicians seem to be more responsible to the voice that they represent. But which flavor is the most effective in general? Singaporean? Scandinavian? Japanese? European? And can this system be improved? Could it be possibly reduced to a five-party parliamentary system? Could representation be clearer so that more people can participate in the political process?

It is also interesting to note that although China has only one party—the Communist Party—the National People's Congress acts as its elected parliament. To be able to manage the most populous nation on earth, with 1.3 billion people, is no easy task. In fact, it is indeed a political miracle that it isn't experiencing any civil war. Its worst problem now is with Taiwan and Tibet. In fact, it is so stable, that it is currently attracting more foreign investment than other countries on earth; and its economy is growing at a rate above 9%. If it keeps going at that rate, it will overtake U.S. GDP a few decades from now. China is also the only living and longest-running continuous civilization since ancient times. Can we learn from their fluid type of governance?

We do know that even with an effective system, the progress of a nation is limited to the type of human resources that it maintains. South East Asians, for instance, know that if they need good athletes for their basketball games, they import blacks or African Americans since they are genetically the best when it comes to sports. Or when it comes to trade and security, they import whites since they are the most driven when it comes to hard work, strength, sales and global marketing. Or when it comes to governance, they often have yellows implementing political and intellectual stability in the region. Of course, when it comes to spirituality, healing and hospitality, the Malay-Chinese are the best in that field.

This means that if you want to increase per capita GDP dramatically, an ASEAN nation must be governed by Chinese or Japanese politicians, partnering with American or European businessmen, and sustained by the easy-going nature of Malays to prevent overheating or breakdowns. The proportion of Chinese to Malays, for example, in the region is generally directly related to the per capita GDP of the individual nations. Indonesia, for instance, having too many Malays and not enough Chinese is very poor and too religious. The Philippines has more active Chinese in their systems and is somewhat better off. Then, Thailand has even more of them and is twice as good as the Philippines. Malaysia, on the other hand, has significantly more, and is more than three times as good. Of course, Singapore, after it broke away from the Federation of Malaysia, became a first-world country in about 30 years, since most Singaporeans are Chinese. Of course, the Singaporeans are now drinking recycled sewage water because they don't have enough Malays to lessen their intellectualism. Therefore, to be able to speed up "progress" means to be able to attract or to import the appropriate human capital, and to manage them in the right proportions, not just to be able to change the political system.

(Incidentally, the word *Tao* in Filipino means "people." And in this human resource case, it means "People Power." Per capita GDP though is a weak barometer of optimal well-being since the figures are not equivalent across the nations. Some currencies, for instance, such as the Chinese Yuan, are undervalued. The cost of living also varies greatly across the continents. One year's rent, for example, in a tiny apartment, in New York City, could *buy* you a fully-owned 3-bedroom house and a garden, in a suburb in South East Asia. Needs and wants also vary among the races. A white man, for instance, who has three cars, may still be discontented. But a Singaporean who takes trains and cabs might be very satisfied already. Indeed, when economists think that all people have unlimited wants, they are only showing off their incompetence. Therefore, someone living in a country with a high per capita GDP may sometimes be even worse off than someone residing in a low-GDP nation. Indeed, many economic indicators are very misleading.)

It is interesting to note that racial types play a significant role in political systems and people management. Super-egotistic Asians, for example, are *sensitive*, and they can't take criticism as well as iditistic Westerners, which is why democracy works in the West and autocracy tends to work better in the East. Center-Left types, such as the Singaporeans, Japanese and blonde Scandinavians, tend to be more honest and united, but they are very boring and lifeless. On the other hand, people on the Right such as the Malays and Africans are disunited and dishonest, but they are full of life. It is then great to do business with the Center-Left, but fun is not their forte. Leisure time is fun with the Center-Right, but doing business with them is painful. Indeed, the 2003 Transparency International global corruption report reinforces this by showing that the left tends to be more honest and tight, while the right tends to be dishonest and loose. Out of a possible score of 10, (Chinese) Singapore fared 9.4, (Malay-Chinese) Malaysia scored 5.2, and (Malay) Indonesia crawled at 1.9. It may appear ironic to observe that militant types are honest and shamanistic types are dishonest, but knowing that the truth hurts and that fantasy heals in the short-run makes you realize that it actually makes sense. As time passes though, the opposite is true: fantasy begins to destroy and reality starts to move you forward. Therefore, to achieve the right balance between the two is an artful skill that managers have to practice. (Take note that the definition of "honesty" in this case comes from the left perspective. "Honesty" defined from the Right will reverse the figures like a mirror. Indeed, the supposedly "honest" West is cunningly deceitful, while the "dishonesty" of the East is emotional and obvious.)

Of course, if you need business drive or psychopathic bloody militancy, you'll have to go with the Britons and their ancestors in America and Australia—the far-right-far-left bipolars. The word "bloody" is commonly attached to their colloquial conversations since their mouths and hands are full of blood, cleansed with their Holy Grail, the cup of Hannibal Lecter. In fact, Californians recently elected Arnold Schwarzenegger as governor, even if he is known to treat women like pieces of meat, since the Californian psyche loves the idea of being a

Terminator. That's probably why the *Silence of the Lambs*, starring Sir Anthony Hopkins as the cannibalistic psychiatrist Dr. Hannibal Lecter, won five major Oscar awards in 1992, including Best Picture. Hollywood folks who voted for it can relate well to it since they believe, just as many Westerners do, that the only way to progress is to "eat" other people alive, since if you are kind, you'll be left behind or "eaten" by someone higher in the food chain. (Of course, Hollywood is not the only place where people "eat" each other to stay ahead of the pack. Take a trip to Vatican City and you will observe cardinals there stabbing each other at the back and "eating" each other, as attested by my American friend who lived there for a while.)

Take note though that there are many types of Americans and they are all very "different." Scandinavian Americans, for instance, in Minnesota or North Dakota are very kind and honest, but an Italian American in the East Coast may eat you alive. An African American in Detroit or Chicago may mug you for shallow reasons, but they can't scare you as much as the "rational" white Lecter-type Americans who can you skin you alive or plot against you because of the color of your skin. This is why the ease of recovering from mental illness and its severity varies according to race, just as breast cancer is least likely among Asians.

In addition, Arab-whites such as the Italians and Spaniards, just like their brunette colleagues in the U.K, tend to be militant types. Bull fights, gladiator matches, rugby and American football are their pastimes, but they can also be very artistic and fashionable since the far left oscillates to the far right. This is why in producing an effective team, different types are necessary to move forward, since the political system by itself is not sufficient enough to produce positive results. The Rights are the creative initiators; the Lefts are the drivers who take action and the Center moderates and caps everything. Indeed, it is like a five-man basketball team with two *guards*, two *forwards* and a *center*. And if managers and politicians can determine the right mix for their members, the organization or country becomes more productive. Take note that it doesn't make anyone necessarily "good" or "evil." It simply considers everyone as part of a spectrum, with different roles to play.

Moreover, in South East Asia, children are taught to call older people using titles, whether they deserve respect or not. Filipinos use titles such as *kuya* or "elder brother" and *ate* or "elder sister" in the Philippines, but other Asians also use similar titles in their local language. In other words, a child, who is a complete stranger to me, might call me *Kuya* Rodney. Then, when addressing someone who is older, you must use the word *opo*, which roughly means, "yes, sir" or "yes, ma'am." This is typical of shamanistic types, which is why a democratic government that receives a lot of media and people-heckling fails to work with them. It then attracts only the iditistic types who end up governing the country miserably because they are the ones who can take the criticism. On the other hand, Westerners, who are more working-class equalitarian-types, do not have such practices, and they will freely call each other by first name only, except perhaps in the church, where they may address each other as "brother" or "sister."

If you also give a hardworking Westerner a tractor after farming manually all his life, he will use the extra time that he has to become more productive after using the time-saving machine. On the other hand, if you give a tractor to a Malay, accustomed to manual farming, he will probably use the extra time that he saves on sleeping. It's not that Malays are lazy or indolent. It's really because shamanistic types are that way. Americans may also prefer to do their own plumbing, paint their own house, fix their own car and perform other types of manual labor proudly. But South East Asians will shun such practices unless the task at hand is something of religious significance. That's why you can motivate them to build nice temples, but not in other things. They will house the gods and their foreign guests lavishly, but they themselves will live very simply. You have to attach some type of spiritual worth to the task at hand, highlighting the divine mission and vision, to make them more productive. It is not like Westerners where the lure of nice cars, houses and lots of money will make them work harder. It is indeed beneficial to know what motivates different types of people in order to manage them more effectively and efficiently.

Another limitation to even the best political system is the factors or events that occur in the external environment. For instance, the condition of neighboring countries or even the cycles of nature may seem far away, but it will definitely affect your country sooner or later. This certainly occurred during the prime of Angkor, Cambodia, between the 10th and 12th century when it was the largest city in the world at that time, populated with about one million people. While she flourished in the arts and sciences using center-left politics, her far-away neighbors up north were extremely miserable. And this misery led to the rise of Genghis Khan and the Mongolian Empire. It became the largest land empire in history, stretching from Hungary in the west to Korea in the east. And since it occupied parts of Thailand also in the south east, the Thais, which used to be friendly toward the Cambodians, became psychotic. And eventually, the Thais became crazy enough to attack Angkor and to destroy the city almost completely. This led to the downfall of the ego-superego period in Cambodia. Indeed, as I explained in my second book, the five faces of society rotate and rise, turn by turn, moving from id to ego-superego in a cyclical pattern. The best thing a politician can do is to minimize the effects of the cycles and to move away from indifference, but somehow, the rotation still takes place to a certain degree. It cannot be completely eliminated. And with today's highly connected world, the domino effect moves very quickly.

Another factor that influences the effectiveness of any political system is the functional education level of the populace. The educational system today is too theoretical. It is easy to find students who think that subjects such as algebra, geometry, history, philosophy, literature and the like, are all "useless" topics. The reason is simple; they don't have any hands-on projects to apply it with. Indeed, a school that directs more real-world activities to its students is more effective than predominantly lecture-type classes, since doing and applying is more effective than simply listening or reading passively, without even having a clue to what it's used for. And in the end, a more functionally-educated public will move in the right direction by themselves *automatically*, minimizing the need for guidance, supervision and management from politicians.

Moreover, would it also be possible to ease the burden of the judicial system? Is it sensible to establish a 5-member judge and jury composed of graduating college students taking 5 different courses, not just law or philosophy, to handle real cases, or at least the minor ones? Can the youth be more effective than old Two-Face? And do women make better judges than men? We do know that women use two brains and men usually concentrate on only one, with no counterbalance. However, using two brains can lead to indecision, which is usually evident in fickle-minded women such as Gloria Macapagal-Arroyo who couldn't decide on whether to run for re-election or not. Indeed, a split mind seems to be more evident in women than men.

In the end, in today's internet-driven, satellite-connected world, the voice of the Five Faces of Society is more audible then ever. Theoretically, a politician is a moderator who harmonizes the conflicting interests of all the sectors in society, but in many countries today, the interests of the politicians are served first because the people or the *Tao* are slow to act in moving their government toward the direction that they want. We do know that the political system will eventually evolve to a more harmonic structure, even if my statements may not be "politically correct," because in the end, the flattery of political correctness is shallow and is a con-man's shamanistic strategy. The *Tao* eventually knows itself better and moves toward what's best for it in the long run. It will, in due course, minimize the friction of politics and disharmony, because that is a key principle of success. The more it thinks in terms of relativity, instead of absolutes, the more it moves toward the "speed of light." Indeed, those who prefer to keep fighting in the political arena will be left behind by the relativists who say that there's no such thing as "right" or "wrong," but only interests that must be harmonized.

In Chapter 8, I will be discussing *world development.* What is the future of the United Nations without a true military arm to enforce its resolutions? The U.S. intimidated it, as usual, during the Iraq-war dispute, but France, Russia and Germany effectively "vetoed" a U.S.-proposed resolution to use military force on Iraq. The U.S. went to war anyway, against the will of the UN, arrogantly thinking that they could go at it alone. But in the latter part of the year, U.S. casualties mounted and they begged the UN for military assistance. So the U.S. proposed a resolution that would supposedly add UN military forces to help the U.S. in Iraq. At first, the early versions of this resolution were rejected again by France, Russia and Germany, but after several American revisions to appease the UN's criticism of the U.S. puppet government in Iraq, the UN Security Council eventually used the paradox-poker-face strategy to confuse the Americans. In other words, they voted for the resolution, but they didn't send any troops, which essentially made the resolution useless.

Truly, politicians have to be good actors and poker players to be able to get what they want, and this is their usual strategy. For example, Bush is known to pledge aid to developing economies, but he doesn't fulfill it. The U.S. government is also known to commonly use semantics to fool everybody. When they had military bases, for example, in the Philippines, they called their rent payment to the Philippine government as "economic aid," but the Philippines called it "rent." The U.S. also calls bribes and attempts at economic sabotage as "economic aid" for third-world countries. And just recently, Bush wanted his loan package to Iraq to be considered as "grants" rather than "loans" to prevent the public from accusing him of wanting Iraq's oil, construction contracts and money. So the U.S. government outdid themselves again with their Oscar-winning performance, pretending to dispute the loans, worth billions of dollars, to turn them into "grants."

Thus, as American politics tries to deceive others around the world, their effectiveness in doing so diminishes over time. Victimized nations begin to gather together and develop their own strategies. And in the long run, the id cannot hold on forever, since the id is only effective in the short run. Other countries develop better game-theory skills and evolve into more clever players after forsaking Mr. Nice Guy. But unlike the precept called Occum's Razor, it is usually not the simplest that's most correct, nor is the simplest the most fit to survive. Actually, given a spectrum of opinions, the most correct one is usually somewhere in the middle; hence it survives the longest. For instance, in the beginning of the year, Bush was confident in bullying North Korea into submission, threatening it like Iraq; but as the months passed, Bush bowed down slowly to North Korea, gradually moving toward what Korea wants—a written peace agreement. But up to now, Bush still can't get what he wants from North Korea.

How then will the different nations be organized in the future? Will there ever be a one-world government? Who will lead the world? We do know that the one in the mid to far left cannot see the people on the right, and vice versa. But the ones at center left and right has the vantage point that can see everything, even the far left and right. Remember though, that humans share a common ancestry with monkeys and fish, so you really can't expect too much from our world leaders. Every sector of society also cannot be pleased completely. That's why true consensus is an asymptote that will never be achieved entirely, although you can come close to it through proper management.

In the end, we do know that the world is more connected now than ever. A problem in one part of the world affects the other parts as well. And in order to manage these problems, the one-world body needs to be effective and efficient. Hence, because of this need, eventually we know that it will be gradually filled to a certain degree, if the earth manages to survive.

In Chapter 9, I will be discussing *space development*. In 2003, during the Chinese New Year, the Columbia space shuttle exploded into what looked like, three major pieces, falling down to earth like three balls of fire or shooting stars. It appeared like Sony's Columbia Tri-star constellation. Is this an omen for the future of NASA? China is planning manned missions to space. Japan is sending a probe to Mars. Will the Asians catch up? The European space agency is also sending probes to Mars, is it all worth it? Can space development actually be hastened by solving earthly problems first? If conflicts or wars are minimized on earth and more people become healthier, wouldn't more people be able to contribute to space development?

Space development is indeed an expensive *fantasy*. Japanese anime delves in this because of their overcrowded psyche. We do know that a Star-Trek-like voyage is impossible since even the nearest star is millions of lifetimes away. Should it be limited to commercial and security purposes only?

In the end, we know that people can't survive in space in the long-run, at least for the next few millennia. Space stations will be eventually destroyed by meteors. And even if man oxygenates Mars and transforms its terrain and atmosphere, its gravitational force is not compatible with the human body, and staying in Mars for the long-term will lead to sickness and death. Traveling also to a nearby star where other planets may reside is not possible; at least not in the near or even distant future. Even if a wormhole can connect our solar system to another one in the distant universe, man cannot survive through a black hole. His molecules will be torn apart. He would have to be in another form to pass through it. But that would already be in the realm of "spiritual" theory. Clearly, we have reached the limit already, so we all have to love Mother Earth because we don't have any other home to go to.

Actually, it may be possible one day to produce human clones that are compatible with Martian gravity. And it may also be possible to completely map and duplicate the contents of someone's mind, just like copying the contents of your hard-disk drive to a CD. Then you could transmit your mind's data via radio waves into a Martian clone in a Martian cloning base. That would theoretically allow space travel and perpetual life. Perhaps that is also how other theoretical beings, which hypothetically evolved before us, cloned themselves and produced humans. But that is too theoretical and far ahead to even think about. Let's be practical here!

Just recently though, when I was about to wrap up this essay, someone noticed some dirt on my back while my shirt was off for a dip. But when I looked at the mirror, I noticed three circular dots, shaped like an inverted V on my back. It looked just like the Middle Way symbol on my second book, but I never noticed it before since it was on my back. It is probably just a coincidental birth mark, but then I remembered Steven Spielberg's *Taken* mini-series. I saw two episodes of that alien-abduction sci-fi series a few weeks ago through the *Star Movies* channel in the Philippines, where it was broadcasted in the 3rd quarter of 2003. It showed different schizophrenic cases of alleged alien abductions, and there was a case of a little girl who also had three little dots shaped like a V supposedly branded on her by aliens as their *signature*. Certainly the coincidence is very outlandish, but

I'd like to stick to the conservative viewpoint and think of my experience from the perspective of medicine. While it may be possible that my weird experience and hence, my books, are a product of alien intervention with their out-of-this-world message, I prefer to conservatively think of it as a product of illness, with the struggle to cure it, attempting to make the world a better place through my own intelligence as a "legal alien" or transient, from an INS perspective, learning from my observations about various types of humans as I travel. Indeed, that's probably how space aliens—if they are truly around—managed to evolve from their odd appearance, ailments, and social problems. And that's why these theoretical beings want to lend their experience to us, so that we can also evolve into happier and healthier entities.

This experience reminds me of the Hindu god Ganesha. I used to think that Hindu gods are weird and horrifying, especially since Ganesha has the head of an elephant and a human body. But then, I spent some time trying to find out why they look so strange, and all of a sudden, it wasn't so grotesquely funny anymore. Hindu myth narrates the tale of a goddess married to Siva, who leaves her alone while he journeys abroad. To comfort her loneliness, she produces a clone boy using the dirt from her body and calls him Ganesha. Then she asks the boy to guard the front door while she dips her naked body for a bath. Eventually, Siva arrives at the door, back from his trip, wondering why it was locked, and demanding that it be opened. But Ganesha refuses to let him in, proclaiming that he was protecting his mother. So Siva became even more furious, since he would certainly recognize his own son if he had one. Finally, Siva barges in, beheads Ganesha, and throws his head toward the mountains. When his shocked mother saw the aftermath of the situation, she explains to Siva that she cloned the boy from her own body using some dirt, and she demands that his life be restored. But when Siva tries to search for the boy's head, he fails to find it; and instead, he beheads an elephant and replaces it on Ganesha's body, which revives the boy; although he now has an elephant head. Realizing that Siva's destructive personality is really the way things are, she simply accepts Ganesha's fate before more harm takes place, harnessing her motherly instinct to continue loving her living boy, who now looks like the *Elephant Man* (1980), rescued by Sir Anthony Hopkins, playing his doctor.

The Indians, now of course, deify this special *Star Child*—as Sir Arthur Clarke calls it in *2001: A Space Odyssey*—believing that one day, they may be awarded their freedom from the Earth Correctional Facility, since Ganesha is said to be the protector of thresholds such as doors, bridges and ports, holding the "keys" to "heaven" and "hell."

The Indians may also eventually join China, which launched a successful manned mission to space last October 15, 2003. Japan and Korea may follow, and the future of mankind may be headed toward the stars. But while we are here on earth, we must remember how Dole shunned their farmers, whose children were born like pineapple-like octopi after they absorbed pesticides forced on them by their employers. Or we must remind ourselves of the Vietnamese children who were victims of America's Agent Orange. Or Americans themselves must commit to memory how their own government experimented on them like guinea pigs, using bio-chem weapons on Americans in Project 11. And we must keep in mind the numerous other cases of Siva-like entities on earth, to manage our destiny.

Finally, in Chapter 10, I conclude everything and summarize my findings. And as Mr. Spock, in the usual Vulcan Star-Trek Greeting would say, "Live Long and Prosper." This is the very same "hello" and "goodbye" of Asia-Pacific people in words like *Aloha, Sawadee, Auyobon* and *Mabuhay*! In the end, we will all finally "graduate" from our earthly existence, and one day we'll be able to tell our "warden," "Beam me, up, Scottie!"

Convergence Axiom

Given a spectrum of opinions, the most correct one is usually somewhere in the middle.